GUIDELINES FOR REVALIDATING A PROCESS HAZARD ANALYSIS

This book is one in a series of process safety guidelines and concept books published by the Center for Chemical Process Safety (CCPS). Refer to www. wiley.com/go/ccps for full list of titles in this series.

GUIDELINES FOR REVALIDATING A PROCESS HAZARD ANALYSIS

Second Edition

WILEY

Registered Office
John Wiley & Sons, Inc., 111 River Street, Hoboken, NJ 07030, USA

Editorial Office
111 River Street, Hoboken, NJ 07030, USA

For details of our global editorial offices, customer services, and more information about Wiley products visit us at www.wiley.com.

Wiley also publishes its books in a variety of electronic formats and by print-on-demand. Some content that appears in standard print versions of this book may not be available in other formats.

Library of Congress Cataloging-in-Publication Data Applied for:

Hardback ISBN: 9781119643609

Cover Images: Silhouette, oil refinery © manyx31/iStock.com;
Stainless steel © Creativ Studio Heinemann/Getty Images;
Dow Chemical Operations, Stade, Germany/Courtesy of The Dow Chemical Company

SKY10036755_101222

TABLE OF CONTENTS

Contents

LIST OF TABLES

LIST OF FIGURES

ACRONYMS AND ABBREVIATIONS

AIChE	American Institute of Chemical Engineers
ALARP	As Low As Reasonably Practicable
ANSI	American National Standards Institute
API	American Petroleum Institute
ASME	American Society of Mechanical Engineers
ATEX	ATmospheres EXplosible
BLEVE	Boiling Liquid Expanding Vapor Explosion
BMS	Burner Management System
BPCS	Basic Process Control System
CCPS	Center for Chemical Process Safety
CIA	Chemical Industry Association
CMMS	Computerized Maintenance Management System
DCS	Distributed Control System
DHA	Dust Hazard Analysis
DMR	Damage Mechanism Review
EHS	Environment, Health, and Safety
EPA	Environmental Protection Agency
ETA	Event Tree Analysis
FMEA	Failure Modes and Effects Analysis
FTA	Fault Tree Analysis
HAZOP	Hazard and Operability
HCA	Hierarchy of Hazard Controls Analysis
HHC	Highly Hazardous Chemical
HIRA	Hazard Identification and Risk Analysis
HRA	Human Reliability Analysis
ICI	Imperial Chemical Industries
IEC	International Electrotechnical Commission
IEF	Initiating Event Frequency
IPL	Independent Protection Layer
ITPM	Inspection, Test, and Preventive Maintenance
LOPA	Layer of Protection Analysis
MAWP	Maximum Allowable Working Pressure

MCC	Motor Control Center
MI	Mechanical Integrity
MOC	Management of Change
MooC	Management of Organizational Change (See OCM)
NFPA	National Fire Protection Association
OBRA	Occupied Building Risk Assessment
OCM	Organizational Change Management
OR	Operational Readiness
OSHA	Occupational Safety and Health Administration
P&IDs	Piping and Instrumentation Diagrams
PFD/PFOD	Probability of Failure on Demand
PHA	Process Hazard Analysis
PPE	Personal Protective Equipment
PLC	Programmable Logic Controller
PSI	Process Safety Information
PSID	Process Safety Incident Database
PSM	Process Safety Management
PSSR	Pre-Startup Safety Review
QA	Quality Assurance
QRA	Quantitative Risk Analysis
RAGAGEP	Recognized and Generally Accepted Good Engineering Practice
RBPS	Risk Based Process Safety
RMP	Risk Management Plan
RP	Recommended Practice
SCBA	Self-Contained Breathing Apparatus
SDS	Safety Data Sheet
SIL	Safety Integrity Level
SIS	Safety Instrumented System
SME	Subject Matter Expert

GLOSSARY AND NOMENCLATURE

This Glossary and Nomenclature section contains process safety terms unique to this Center for Chemical Process Safety (CCPS) publication. The CCPS process safety terms in this publication are current at the time of issue. For other CCPS process safety terms and updates to these terms, please refer to the "CCPS Process Safety Glossary" [1].

Complementary Analysis/Analyses: Beyond the core analysis, this includes any additional analyses on specific topics required by policy or regulation to complete the minimum requirements for a process hazard analysis (PHA). Complementary analyses often address topics such as human factors and facility siting, and range in complexity from simple checklist analyses to sophisticated computer modeling. Like the core analysis, any complementary analyses must be included in the scope of a revalidation.

Core Methodology/Core Analysis: This is the primary method used to identify, evaluate, and document process hazards and their consequences and risk controls. Although other PHA methods are available, one of the following three techniques is most commonly used: Hazard and Operability (HAZOP) Study, What-If/Checklist, or Failure Modes and Effects Analysis (FMEA).

Facilitator/Leader/Study Leader: Universally, PHA revalidation teams must include at least one person knowledgeable in the hazard analysis methodology being used. The terms "facilitator," "leader," and "study leader" are used interchangeably in this book, as they are in general practice, but the person's actual role is a blend of both facilitator and study leader. As facilitator, they should engage all other team members in the correct application of the chosen hazard analysis technique and make the work process easier. They should also ensure that the PHA team follows the relevant rules and guidelines specified for the study. Simultaneously, they should lead the team to complete the revalidation efficiently and accurately without stifling the team's imagination and curiosity as they try to identify hazards of the process.

Hazard Identification and Risk Analysis (HIRA): This collective term encompasses all activities involved in identifying hazards and evaluating risk at facilities, throughout their life cycle, to make certain that risks to employees, the public, or the environment are consistently controlled within the risk tolerance of the organization. PHA revalidation, as described in this book, is also generally applicable to the revalidation of HIRAs.

Incident: Use of the single term "incident" throughout this book is intended to include of the incident definitions listed in the CCPS glossary definition. These include "catastrophic incident," "incident," "near-miss incident," and "process safety incident/event."

Process Hazard Analysis (PHA): This review is an organized effort to identify and evaluate hazards associated with processes and operations to enable their control. Normally, the analysis involves use of qualitative techniques to identify and assess the significance of hazards. Reviewers judge risks and develop appropriate recommendations. Occasionally, quantitative methods are used to help prioritize risk reduction. In this book, the term "PHA" encompasses all the activities involved in creating and revalidating the resulting PHA document.

Process Safety Management (PSM) System: A management system that focuses on prevention of, preparedness for, mitigation of, response to, and restoration from catastrophic releases of chemicals or energy from a process associated with a facility. In this book, the term "PSM" broadly encompasses management systems anywhere in the world with similar intent, regardless of their official name, whether required by regulatory authorities or by organizational policies. PSM issues related to a specific jurisdiction, such as the United States Occupational Safety and Health Administration (OSHA) PSM, are noted in this book.

Semi-Quantitative Risk Analysis: This evaluation uses a risk matrix with a numeric frequency scale and a descriptive, qualitative consequence scale. Any risk analysis based on such a matrix is inherently semi-quantitative (part numeric, part descriptive). This terminology is sometimes incorrectly used to describe a simplified, but fully numeric, QRA.

Simplified Quantitative Risk Analysis/Simplified Process Risk Assessment: Quantitative risk analysis (QRA) involves the systematic development of numerical estimates of the expected frequency and/or severity of potential consequences associated with a facility or operation. The estimation process may be based on conservative rules and order-of-magnitude data to simplify the analysis. Simplified QRA techniques, such as Layer of Protection Analysis (LOPA), are typically used to identify whether the risks of evaluated loss scenarios are managed to tolerable levels.

Supplemental Risk Assessment: This includes any additional analyses used to improve the quality and consistency of risk judgments in a PHA. Common techniques range from (1) simple categorization on a risk matrix to (2) simplified risk analysis techniques such as LOPA and bow tie analysis to (3) detailed QRAs. Supplemental risk assessments are typically included in the scope of a revalidation to be consistent with the requirements of the facility owner and applicable regulations.

ACKNOWLEDGMENTS

The American Institute of Chemical Engineers (AIChE) and the Center for Chemical Process Safety (CCPS) express their gratitude to all members of the *Revalidating Process Hazard Analyses Second Edition* subcommittee and their CCPS member companies for their generous efforts and technical contributions in the preparation of this book.

The chair of the subcommittee was Laura Acevedo Turci from 3M. Warren Greenfield was the CCPS consultant. The subcommittee also included the following people who participated in the writing of this *Guidelines* book:

Alok Khandelwal	Shell
Curtis Clements	Chemours
Jennifer Bitz	CCPS
Jordi Costa Sala	Celanese
Lisa Parker	Braskem
Manuel Herce	DuPont
Marty Timm	Praxair - Linde - Retired
Michael Hazzan	AcuTech Consulting
Mohammad Baig	Bayer

CCPS especially wishes to acknowledge the contributions of the principal writers from ABSG Consulting Inc. (ABS Consulting):

Derek Bergeron
Bill Bradshaw
Don Lorenzo

The writers wish to thank the following ABS Consulting personnel for their technical contributions and reviews: Tom Williams and David Whittle provided technical review of the book. Lynn Purcell edited the manuscript. Susan Hagemeyer helped prepare the final manuscript for publication.

Before publication, all CCPS books are subjected to a thorough peer-review process. CCPS also gratefully acknowledges the thoughtful comments and suggestions of the following peer reviewers. Their work enhanced the accuracy, clarity, and usefulness of this *Guidelines* book. Although the peer reviewers have provided many constructive comments and suggestions, they were not asked to endorse this book and were not shown the final manuscript before its release.

Ahmad Hafiz Zeeshan	Tronox Australia
Allison Knight	3M
Bernard Groce	Calumet Specialty
Dan Sorin	Novachem
Denise Chastain-Knight	exida
Don Connolley	aeSolutions
Felix Azenwi Fru	National Grid
Gill Sigmon	AdvanSix
James Martin	Corteva™ Agriscience
Kristine M. Cheng	Chevron
Lisa Long	OSHA
Matias Massello	Process Improvement Institute
Michell LaFond	Dow
Michelle Brown	FMC
Muddassir Penkar	Heartland Petrochemical Complex Interpipeline
Nick Ashley	W.R. Grace and Company
Palaniappan Chidambaram	DSS
Rizal Harris Wong	Petronas
Robert Feuerhake	Lanxess
Ruel Ruiz	3M
Scott E. Swanson	Intel
Tekin Kunt	PSRG

PREFACE

The American Institute of Chemical Engineers (AIChE) has helped chemical plants, petrochemical plants, and refineries address the issues of process safety and loss control since 1967.

The Center for Chemical Process Safety (CCPS), a directorate of AIChE, was established in 1985 to develop and share technical information for use in the prevention of major chemical accidents. CCPS is supported by a diverse group of industrial sponsors in the chemical process industry and related industries who provide the necessary funding and professional guidance for its projects. The CCPS Technical Steering Committee and the technical subcommittees oversee individual projects selected by the CCPS. Professional representatives from sponsoring companies staff the subcommittees and a member of the CCPS staff coordinates their activities.

Since its founding, CCPS has published many volumes in its "Guidelines" series and in smaller "Concept" texts. Although most CCPS books are written for engineers in plant design and operations and address scientific techniques and engineering practices, several guidelines cover subjects related to chemical process safety management. A successful process safety program relies upon committed managers at all levels of a company, who view process safety as an integral part of overall business management and act accordingly.

This book is an update of the 2001 book "Revalidating Process Hazard Analyses". In 21 years since the publication of the first edition, PHA and PHA revalidation have evolved, and the methods for conducting them have changed and improved. This updated edition addresses the complementary analyses and supplemental risk assessments that often are incorporated into PHAs. It reflects the knowledge gained by the CCPS, writers, and the project committee members, who collectively, have more than 100 years of practical experience in conducting and revalidating PHAs.

This book was written to enhance the knowledge and provide insight and guidance to those who are involved with the planning and execution of PHAs and are responsible for updating and revalidating PHAs. Multiple audiences should find this book useful. Those who manage the PHA program will find guidance for resources and timing. Those who must determine the type of revalidation necessary will find tools for analyzing the past PHA and related studies. PHA teams will find checklists and other tools in the appendices for use during the study.

DEDICATION

Updating and Revalidating Process Hazard Analyses

Is dedicated to

Walt Frank

Walt Frank is a process safety professional with a wide range of experience and expertise that is helping to improve process safety worldwide.

From his early entry into process safety at DuPont through consulting at ABS and later as President of Frank Risk solutions, Walt has always been dedicated to sharing process safety knowledge and expertise through his many contributions to both NFPA and CCPS.

Walt was a primary writer for the first edition of the Revalidating Process Hazard Analyses. He has also been a key contributor to several other CCPS books. In addition, he is the current chair of the CCPSC Exam committee and is a key contributor to the CCPS Golden Rules for Process Safety project.

Walt is often the driving force that helps teams focus on the key process safety aspects requiring development and presentation in a clear and precise manner. He has a keen eye for details and sets similar expectations for his colleagues and team members. He is persistent in pursuing perfection.

Walt is a Registered Professional Engineer, AIChE Fellow, CCPS Fellow, Certified Process Safety Professional (CCPSC) and a CCPS Emeritus. CCPS is proud to dedicate this book to Walt in recognition for all he has given and continues to give, to the process safety community.

Anil Gokhale & Warren Greenfield

INTRODUCTION

OBJECTIVE OF THIS BOOK

For years, formal process hazard analyses (PHAs) have been performed on processes handling hazardous materials, and most companies have included requirements for the conduct of PHAs in their process safety management (PSM) programs.

Most company PSM programs also include requirements to revalidate the PHA periodically. Such requirements are intended to ensure the PHA for a process is complete, up to date, and thoroughly documented, considering the changes and incidents that have occurred in the process and the operating experience gained since the prior PHA was conducted. In jurisdictions where PHAs are required by regulation, periodic revalidation is usually required to maintain regulatory compliance.

Because the efforts to revalidate the PHA can involve a significant investment of time and resources, it is important that the revalidation effort be thoughtfully organized, conducted, and documented. This book addresses the need, identified by the CCPS, for supplemental guidance with respect to considerations unique to the PHA revalidation task.

SCOPE OF THIS BOOK

This book provides an organized approach, and several tools, to revalidate a PHA. However, it does not provide a "one-size-fits-all" approach for revalidation. Neither is it intended to provide the "one true solution" to the task. Readers and companies should develop procedures based upon an evaluation of their specific requirements. Although the concepts in this book could be adapted for revalidation of other studies (e.g., security analysis and financial risk analysis), the book was written exclusively with the intent of revalidating a PHA and the documentation/studies directly associated with a PHA. Revalidation of other stand-alone analyses is outside the specific scope of this book.

This book assumes that the reader is familiar with the conduct of a PHA. While some introductory material on general PHA activities, methodologies, and risk assessment is provided in Chapter 1, this book is not intended to be an instructional text covering the general conduct of a PHA or application of supplemental risk assessment techniques (such as Layer of Protection Analysis [LOPA] or bow tie analysis).

This book is a supplement to the CCPS book *Guidelines for Hazard Evaluation Procedures* [2, p. 8] and is premised on the same concept stated in the book:

> *This book does not contain a complete program for managing the risk of chemical operations, nor does it give specific advice on how to establish a hazard analysis program for a facility or an organization. However, it does provide some insights that should be considered when making risk management decisions and designing risk management programs.*

This book outlines a demonstrated, common-sense approach for resource-effective PHA revalidation. This approach first examines a number of factors, such as PHA requirements, the quality of the prior PHA, and the operating experience gained since the prior PHA (including changes and incidents that have occurred). A revalidation approach is developed based upon this input. The revalidation concept described in this second edition has been updated based on experience and knowledge gained since the first edition was published over 20 years ago.

HOW TO USE THIS GUIDELINES BOOK

To use this book effectively for PHA revalidation, it is helpful to follow the chapters in sequence. The book progresses through each key/major step in the process of revalidating a PHA.

Chapter 1 – Overview of the PHA Revalidation Process explains the role of PHA and PHA revalidation in understanding and managing risk. It describes two revalidation approaches and the typical analytical tools used in revalidation activities.

Chapter 2 – PHA Revalidation Requirements explains the external and internal requirements that must be satisfied by a periodic PHA revalidation.

Chapter 3 – Evaluating the Prior PHA explains how to identify deficiencies in the prior PHA (with respect to current requirements) and to evaluate its usability in revalidation activities.

Chapter 4 – Evaluating Operating Experience Since the Prior PHA explains how to gather and evaluate information (arising from process changes, equipment changes, procedure changes, organizational changes, new research, incident investigations, etc.) that could materially alter the risk judgments documented in the prior PHA.

Chapter 5 – Selecting an Appropriate PHA Revalidation Approach explains how to use the knowledge and insights gained from evaluating the prior PHA and operating experience to select a thorough and efficient revalidation approach.

Chapter 6 – Preparing for PHA Revalidation Meetings explains how to assemble the resources necessary to conduct revalidation meetings and how to document them successfully.

Chapter 7–Conducting PHA Revalidation Meetings explains how to facilitate the use and interaction of resources to achieve the revalidation objectives.

Chapter 8 – Documenting and Following Up on a PHA Revalidation explains how to create a revalidation report likely to meet applicable requirements and how to help ensure elevated risks discovered during the revalidation are communicated and resolved.

Appendices – Sample Checklists are available in the appendices for use in PHA revalidation. Electronic versions of the checklists in the appendices are available at

www.aiche.org/ccps/publications/reval2ed

Password: revalbooked2

1 OVERVIEW OF THE PHA REVALIDATION PROCESS

A process hazard analysis (PHA) is foundational in helping facility management implement and maintain all four of the accident prevention pillars identified within the *Guidelines for Risk Based Process Safety* (RBPS) [3, p. 3]. It may also be required to comply with applicable process safety regulations and internal company requirements. The pillars are listed here and discussed in Section 1.8:

- Commit to process safety
- Understand hazards and risk
- Manage risk
- Learn from experience

Once management has committed the organization to process safety, the next step is to understand what hazards need to be managed. In 2008, the Center for Chemical Process Safety (CCPS) updated and republished its *Guidelines for Hazard Evaluation Procedures* book, which includes the following definition of a hazard [2, p. 51]:

> *A hazard is a physical or chemical characteristic that has the potential for causing harm to people, property, or the environment. Thus, hazard identification involves two key tasks: (1) identification of specific undesirable consequences and (2) identification of material, system, process, and plant characteristics that could produce those consequences.*

The RBPS term "Hazard Identification and Risk Analysis (HIRA)" encompasses the application of a broad range of analytical tools, including those used in a PHA to identify hazards and evaluate risk. A PHA report documents the results from a particular application of HIRA tools intended to meet specific requirements for managing risk in a process. These requirements can be internal (e.g., company policy) and/or external (e.g., regulatory). Over the life of a process, these requirements and their interpretation may change. There may also be external changes, such as community development or rainfall patterns that affect risk. The company's experience with and understanding of the process will increase, and the company will likely make changes to the process itself. Thus, there are also requirements for periodic revalidation of a PHA to incorporate and address such changes.

1.1   WHAT IS A PHA AND WHAT IS A PHA INTENDED TO ACCOMPLISH?

A PHA involves a variety of activities intended to identify, evaluate, and document the hazards in a chemical process. Those hazards may be related to equipment design, normal or transient operating practices and procedures, maintenance practices, human factors, facility siting, external events, or other areas of concern. (The unique hazards associated with transient operations are discussed in the CCPS book *Guidelines for Process Safety During the Transient Operating Mode: Managing Risks During Process Start-ups and Shut-downs* [4].) The resulting PHA report also documents the PHA team's judgments related to the design, operation, and maintenance of a process that could lead to accidental chemical releases, fires, explosions, or other loss events. This information is very valuable in the planning and design of all the other RBPS elements. Most importantly, the PHA report documents the team's analysis of those loss scenarios that may pose an elevated risk to personnel, property, and/or the environment, key safeguards, and any team recommendations for further risk reduction.

As shown in Figure 1-1, PHA activities typically involve the following five steps:

1. Identify hazards (potential for adverse consequences), including characteristics of the following:
    - Process chemicals (e.g., toxicity, reactivity, or flammability)
    - Process equipment (e.g., design specifications, layout, or spacing) affecting the likelihood or severity of adverse consequences
    - Process operations (e.g., potential for high temperature, high pressure, or contamination)
    - Siting (e.g., locations of occupied structures or climate) affecting the likelihood or severity of adverse consequences

2. Identify specific hazardous events (e.g., catastrophic equipment failure, loss of primary containment of hazardous material, uncontrolled energy release) that could result in adverse consequences due to potential causes involving:

   - Equipment performance gaps
   - Human performance gaps
   - External events (e.g., change in process inputs, loss of essential utilities, or natural disaster)

3. Judge risk by evaluating the following:

   - Worst credible case magnitude (severity) of adverse consequences
   - Frequency of events that could result in those consequences (without taking credit for any risk controls)
   - Loss scenario frequencies, taking credit for any relevant risk controls (or safeguards)

4. Decide whether the level of risk is tolerable and/or as low as reasonably practicable (ALARP)

5. If risk is not ALARP, propose risk reduction alternatives (recommendations)

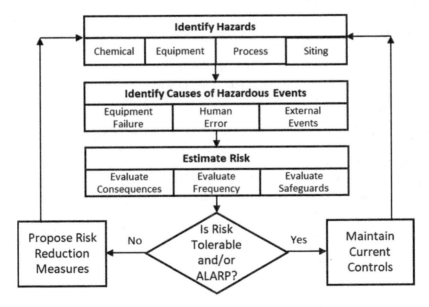

Figure 1-1 Typical PHA Flowchart

1.2   OVERVIEW OF TYPICAL PHA ACTIVITIES

While many factors contribute to a high-quality PHA, proper selection of the PHA core methodology is critical to a successful PHA.

1.2.1   PHA Core Methodology

The PHA is built upon one or more core analytical methodologies. The most commonly used techniques in the chemical/process industries are:

- Hazard and Operability (HAZOP) Study
- What-If/Checklist Analysis
- Failure Modes and Effects Analysis (FMEA)

A detailed discussion and comparison of the PHA core methodologies is beyond the scope of this book. Readers unfamiliar with the various techniques may refer to the CCPS book *Guidelines for Hazard Evaluation Procedures* for guidance on technique selection [2, pp. 99-202]. Descriptions of the more widely used core methodologies have been abridged from the aforementioned book and are provided. Note that any technique used as a core methodology must be applied by a team with all the skills required to perform a PHA.

HAZOP Study. The HAZOP technique is the most often used core methodology. HAZOP is a brainstorming approach involving an interdisciplinary team. The purpose of a HAZOP Study is to review a process or operation carefully and systematically to determine whether deviations from the design or operational intent can lead to undesirable consequences. The team leader, knowledgeable in the methodologies being applied, systematically prompts the team to consider plausible deviations within specific equipment groupings and/or batch steps (commonly called "nodes" or "sections") of the process. Deviations are formed by combining a fixed set of "guide words" (e.g., no, more, less) with the relevant process parameters in that node (e.g., flow, temperature, pressure) and/or the batch step intentions (e.g., adding chemicals in a specified sequence). For example, the guide word "No" combined with the process parameter "Flow" results in the deviation "No Flow." This technique can be used for continuous or batch processes and can be adapted to evaluate written procedures for normal or transient conditions.

For each deviation, the team (1) determines if any consequences of interest would result from credible causes of deviation, (2) identifies engineering and administrative safeguards and controls protecting against the deviation, (3) evaluates risk associated with the deviation, and (4) if necessary, makes recommendations to reduce the likelihood of the deviation or the severity of the consequences.

Similar to the What-If/Checklist technique discussed in this section, a HAZOP is often paired with checklists or other "complementary analyses" to further evaluate specific concerns (e.g., human factors, facility siting).

What-If/Checklist Analysis. The What-If Analysis technique alone, or the Checklist Analysis technique alone can be an appropriate core methodology for some processes. However, in most cases, the two are combined to take advantage of both the creative, brainstorming questioning of the What-If Analysis method and the systematic questioning of the Checklist Analysis method. What-If/Checklist is often used for lower hazard processes, simpler processes, and processes that do not involve chemical reactions, such as storage or transfer systems. What-If/Checklist is also appropriate for processes and activities that do not have rigidly defined material and energy flow paths or that have hazards not directly tied to process parameters such as flow, level, temperature, and pressure. Laboratory and pilot plant operations are often candidates for What-If/Checklist Analysis. This technique is often the first hazard evaluation method performed on a process (e.g., during early design stage of a new process), and as such, it can be a precursor for future, more detailed, studies.

In the What-If portion of the analysis, the team asks questions or voices concerns about possible undesired events. Their concerns (usually phrased in the form of *what if?* questions) are primarily formulated to identify hazardous events or specific loss scenarios that could produce an adverse consequence. There may be no specific pattern or order to these questions, unless the leader systematically directs the team to focus on a particular portion of the process and/or a particular category of questions (process variable deviations, equipment failure modes, operating errors, etc.). The questions can also address any abnormal condition related to the plant or procedure, whether caused by human or equipment performance gaps, or by external events (e.g., floods, high winds, vehicle impacts, dropped objects).

In the checklist portion of the analysis, the team primarily relies on a standard list(s) of questions developed based on experience with similar chemicals, equipment, processes, and/or activities. Frequently, checklists are created by simply organizing information from current relevant codes, standards, and regulations. The standard questions may be modified to address specific concerns about the subject process.

Regardless of the origin of a particular What-If or Checklist Analysis question, the analysis prompts the hazard evaluation team to consider and document potential abnormal situations, their consequences, and the risk controls. The team judges whether the risks are controlled within the organization's tolerance, and if not, makes recommendations for improvement.

FMEA. The FMEA technique is a systematic analysis of single component or functional failure modes and each failure mode's potential effect(s) on a system. For each failure mode, the team (1) determines if any consequences of interest would result, (2) identifies engineering and administrative safeguards and controls protecting against the failure mode, (3) evaluates risk associated with the failure mode, and (4) if necessary, makes recommendations to reduce the likelihood of the failure mode or the severity of the consequences.

The FMEA method is best suited for analysis of mechanical and electronic systems, so it is sometimes selected as the core methodology for processes involving compressors, centrifuges, or highly automated systems. However, FMEA is poorly suited for processes involving reactive chemistry or human interactions where the PHA needs to consider human errors as causes of hazardous events. Therefore, the overwhelming majority of PHA teams use the HAZOP or the What-If/Checklist combination as their core methodology. FMEA is more often applied as a complementary analysis for specific, complex equipment to generate a systematic reference list of components, their failure modes and effects, and prioritized maintenance task lists for increasing equipment reliability.

1.2.2   PHA Complementary Analyses

In a PHA, complementary analyses are used by the PHA team in addition to the core methodology to help provide focus and direction on specific topics. The two most common complementary analyses address human factors and facility siting. These complementary analyses are often reviews of checklists that have been developed by a company or adapted from industry publications and standards. However, complementary analyses can involve much more sophisticated studies such as computational fluid dynamic analysis and modelling of vapor dispersion or finite element analyses of structural response to overpressure and dynamic loads. While the selection, development, and application of these analyses are outside the scope of this book, it is important to note that complementary analyses required by a regulation or company policy must also be revalidated.

Note that different complementary analyses can be used to address the same topic or area of concern. For example, facility siting issues affecting PHA results can be identified and evaluated in multiple ways, including: (1) the consideration of siting during the core analysis, (2) a facility siting checklist review as a complementary analysis, and/or (3) a detailed analysis in accordance with recognized and generally accepted good engineering practices (RAGAGEPs) applicable to a process. For example, the American Petroleum Institute (API) has developed recommended practices for assessing facility siting topics in subject processes [5] [6] [7]. The Chemical Industry Association (CIA) provides guidance for performing occupied building risk assessments (OBRAs) [8]. Such RAGAGEP analyses are often quantitative (either consequence-based or risk-based) and supported by checklists, calculation sheets, and/or models of varying complexity. Company policy may require such assessments, but if they are not explicitly included in PHA requirements, they are not complementary analyses for the purposes of this book.

Examples of complementary analyses are listed in Table 1-1 along with CCPS or other industry publications where additional information can be found.

Table 1-1 Complementary Analysis Examples

Complementary Analysis	Information Source
What-If Analysis*	CCPS book *Guidelines for Hazard Evaluation Procedures* [2, pp. 100-106]
Checklist Analysis*	CCPS book *Guidelines for Hazard Evaluation Procedures* [2, pp. 107-114]
	FM Global Checklists
	Industry publications, guidelines, and recommended practices (RPs)
Facility Siting Study	API RP 752 [5], 753 [6] , 756 [7]
	CCPS book *Guidelines for Siting and Layout of Facilities* [9]
	CIA OBRA [8]
Human Factors Analysis	CCPS book *Guidelines for Preventing Human Error in Process Safety* [10]
Hierarchy of Hazard Controls Analysis (HCA)	CCPS book *Inherently Safer Chemical Processes: A Life Cycle Approach* [11]
Damage Mechanism Review (DMR)	API RP 571 [12]
	API RP 970 [13]
Dust Hazard Analysis (DHA)*	CCPS book *Guidelines for Combustible Dust Hazard Analysis* [14], National Fire Protection Association (NFPA) Standard 652 [15]

can be a core analysis or a complementary analysis in a PHA

1.3   GENERAL RISK ASSESSMENT PRINCIPLES

Developing an understanding of risk requires addressing three specific questions:

1. What can go wrong? (What are the hazards?)
2. What is the [credible] potential impact? (How bad could it be?)
3. How likely is the event to occur? (How often might it happen?)

The PHA team systematically addresses each of these questions for the range of credible loss scenarios. This gives the team a more complete understanding of the risks associated with a process and a basis for judging which, if any, exceed the organization's risk tolerance. If the risk is not tolerable or ALARP, this risk understanding typically leads to recommendations to better control the risk, which are addressed and resolved by facility management.

1.3.1   Risk and Risk Tolerance

The overall risk of a loss scenario is often represented as a function of the consequence **Severity** and **Likelihood**, where **Severity** is "How bad could it be?" In PHAs, severity generally refers to the unmitigated consequences – the maximum credible loss if none of the safeguards worked. **Likelihood** is "How often might the event occur?" Likelihood may refer to either the frequency or probability of an event. However, in PHA risk calculations, the likelihood generally needed is the mitigated frequency of a loss scenario (the frequency of a cause adjusted downward to take credit for applicable safeguards), and it is usually defined as the number of events per year.

To promote consistent risk judgments, PHA procedures rely upon management defining and communicating its risk tolerance to the team before the PHA is initiated. One simple risk tolerance criterion could be an organizational requirement to comply with RAGAGEPs. When PHA teams identify a nonconformance, they recommend corrective action. Many organizations go a step further and develop a risk matrix to communicate their tolerance for any loss scenario falling within a specific range of severity and frequency. PHA teams apply this risk matrix to scenarios identified during the PHA, and if a risk ranking is too high, recommendations are made to reduce risk to tolerable levels.

An example risk matrix is shown in Figure 1-2, and each of its 25 cells is assigned to one of three risk levels – Low, Medium, or High. To be useful in practice, this example matrix would have to be accompanied by the organization's definition of each severity and frequency category, and those definitions could be qualitative or quantitative. A Category 1 frequency might be *"an event never seen in industry,"* or *"an event expected to occur less than once in 100,000 years."* A Category 5 severity might be *"an event expected to adversely affect the public"* or *"a release posing a serious danger to people more than 1 kilometer (0.6 mi) downwind."* In addition, the organization would have to provide guidance on what should be done with scenarios in each risk category. The revalidation team might be required to make recommendation(s) to reduce the risk of scenarios in any of the "High" risk cells, or they might only be required to make recommendations if the risk is not deemed ALARP.

SEVERITY			LIKELIHOOD		
5	Medium	Medium	High	High	High
4	Low	Medium	Medium	High	High
3	Low	Low	Medium	Medium	High
2	Low	Low	Low	Medium	Medium
1	Low	Low	Low	Low	Medium
	1	**2**	**3**	**4**	**5**

Figure 1-2 Example Risk Matrix

1.3.2 Supplemental Risk Assessments

The qualitative risk judgments of PHA teams have proven to be subjective, depending on the knowledge and experience of individual team members. Several supplemental risk assessment tools are available to PHA teams to guide and improve the consistency of their risk judgments. While detailed discussion and comparison of these supplemental risk assessment techniques are beyond the scope of this book, short descriptions of some of the more common tools/methods are provided in this section.

Layer of Protection Analysis (LOPA). LOPA is a simplified quantitative risk analysis method for estimating scenario frequencies using a prescriptive, conservative set of rules. Following those rules, order of magnitude values are assigned to initiating event frequency, probability of independent protection layers (IPLs) failing on demand, and other factors that affect the frequency of the scenario. LOPA is often applied to a selected subset of the individual loss scenarios identified in the core analysis. The purpose of LOPA is to determine whether existing IPLs reduce the risk of a specific scenario to meet the company's risk tolerance criteria. Additional guidance on the application of LOPA, including categorization of initiating event frequencies (IEFs), IPL credits, enabling conditions, and conditional modifiers, is in the CCPS concept book *Layer of Protection Analysis: Simplified Process Risk Assessment* [16] and CCPS books *Guidelines for Initiating Events and Independent Protection Layers in Layer of Protection Analysis* [17] and *Guidelines for Enabling Conditions and Conditional Modifiers in Layer of Protection Analysis* [18].

> **Simplified Quantitative or Semi-quantitative?**
>
> LOPA is a quantitative frequency analysis tool, but the analysis is simplified by standardized rules and order-of-magnitude data. Some organizations use a risk matrix that has a numeric frequency scale and a descriptive, qualitative consequence scale. Any risk analysis based on such a matrix is inherently semi-quantitative (part numeric, part descriptive), even though a quantitative LOPA was used to categorize the loss scenario

Bow Tie Analysis/Bow Tie Barrier Analysis. Like LOPA, a bow tie analysis is applied to specific loss scenarios. However, it is most useful and unique in producing a powerful visual representation (a bow tie diagram) of critical layers of protection that can be readily understood by people at all levels, ranging internally from new operators to senior managers and externally from regulators to members of the public. On the left, a bow tie diagram shows "threats" that have the potential to cause loss of control of the hazard, causing an event of interest (top event) in the middle, resulting in one or more consequences on the right. In a bow tie diagram, the layers of protection against threats (prevention barriers) are separated from the layers of protection against consequences (mitigation barriers). The analysis of both prevention and mitigation barriers can go deeper, identifying and quantifying degradation factors (i.e., potential reductions in effectiveness or reliability) as well as administrative controls to manage the degradation factors.

Thus, like LOPA, it can be used to estimate the frequency of specific loss scenarios. Additional guidance on the application of bow tie barrier analysis can be found in the CCPS concept book *Bow Ties in Risk Management: A Concept Book for Process Safety* [19].

Quantitative Risk Analyses (QRA). QRA is the collective term for a variety of detailed quantitative analysis tools used in risk calculations. In a QRA, the calculated consequences and the calculated frequency of each scenario may be used (separately or combined) to determine individual scenario risks. Individual scenario risks can be further combined to provide an overall picture of cumulative site risks. For example, QRA consequence analysis tools can account for particular release characteristics, such as the density and state (solid, liquid, gas) of the released material, its direction and momentum, its chemical reactions and interaction with humidity in the atmosphere, its interaction with surrounding structures or terrain, the biologic effects on potentially exposed individuals, and so forth. Other consequence analysis tools can calculate detailed fire or explosion effects. Similarly, QRA frequency analysis tools, such as Fault Tree Analysis (FTA), can predict the frequency of system failures from data on the failure rates of specific components. Event Tree Analysis (ETA) and Human Reliability Analysis (HRA) can calculate the likelihood (probability or frequency) of particular loss scenarios. All of these methods require construction of a valid logic model and its reduction to sets of basic causes for which data is available. Failure data is available from sources such as the CCPS book *Guidelines for Process Equipment Reliability Data, with Data Tables* [20].

These QRA frequency analysis techniques are not encumbered by the conservative assumptions embedded in a simplified QRA technique such as LOPA, so they typically produce better estimates of event frequencies. However, they require significantly more analytical and computational effort than the simplified techniques. QRA tools are often used in facility siting studies, in risk analyses of occupied structures, or in the design of Safety Instrumented Systems (SISs). Additional guidance on the application of QRA can be found in the CCPS book *Guidelines for Chemical Process Quantitative Risk Analysis* [21].

1.4   PHA REVALIDATION OBJECTIVES

The primary objective of a PHA revalidation is to produce a revised PHA that adequately identifies and evaluates risk controls for the hazards of the process, *as they are currently understood.*

Reasons why today's understanding of the risks associated with the process might differ from the understanding that existed at the time of the prior PHA include:

- Process changes (including decommissioning of equipment or application of inherently safer design principles) have introduced new hazards or affected existing hazards
- Changes in on-site or off-site facility occupancy or siting have changed the at-risk populations
- External factors, such as the potential for flooding, have changed
- New knowledge is available to better understand the hazards
- Incidents (including events with actual losses or near-miss incidents) in the process or in other similar processes have revealed hazards or specific routes to hazardous events not previously addressed
- Safeguards previously credited in the PHA have been removed, compromised, rendered ineffective, or discredited
- New safeguards have been added
- Applicable codes, standards, or engineering practices have changed
- Risk management systems have been implemented, changed, or discontinued
- Retirements, transfers, or resignations have significantly changed the overall skill levels of the staff

PHAs are an essential element of any process safety management (PSM) system, and they must provide current, accurate information to serve their purpose successfully. Five primary reasons to revalidate PHAs on a periodic basis are:

1. Evaluate the cumulative effect of changes in the process, equipment, or personnel
2. Correct gaps and deficiencies in the prior PHA
3. Incorporate new knowledge and operating experience
4. Comply with frequency requirements per regulations or company policy
5. Comply with content requirements per regulations or company policy

Evaluate the Cumulative Effect of Changes in the Process, Equipment, or Personnel. Process, equipment, and organizational changes must be managed. Most PSM models, such as the 20-element RBPS model developed by the CCPS, include a management of change (MOC) element to ensure that changes are thoroughly evaluated and properly authorized before implementation. However, MOC requests are often evaluated within a rather narrow context (i.e., "What are the potential consequences of what we intend to do right now?"). In cases where processes undergo frequent changes (e.g., a multi-purpose unit), several sequential changes (e.g., a phased equipment installation), or multiple simultaneous changes (e.g., during a unit shutdown/turnaround), the significance of a particular change may not be accurately assessed. In addition, subtle/creeping changes, or the interaction of various changes, may be overlooked. Finally, not all changes may have been captured and evaluated under the MOC program, or temporary changes may not have been reverted to normal service. Thus, PHA revalidation offers an opportunity, on a periodic basis, to perform an integrated evaluation of the cumulative (and potentially interrelated) impact of all of changes, both controlled and uncontrolled.

Correct Gaps and Deficiencies in the Prior PHA. Gaps are errors of omission (i.e., failures to address the established requirements applicable to the prior PHA). Some companies have documented requirements for what a PHA must address. In addition, some regulations, as will be discussed in Chapter 2, are very specific in defining issues that must be addressed during PHAs (e.g., the consideration of human factors). Failure to address human factors, or some other requisite company or regulatory consideration, would be a gap that must be filled during the PHA revalidation.

Deficiencies, on the other hand, are errors in applying the PHA methodology. For example, the prior PHA team may not have consistently (1) traced the effects of loss scenarios to their ultimate consequences, (2) judged the severity or risk of similar scenarios, or (3) documented all the engineering and administrative controls upon which their risk judgments were based. Revalidation teams, often including several or entirely new members, may identify hazards that the prior team did not document.

Incorporate New Knowledge and Operating Experience. PHA revalidation teams may have access to information that the prior PHA team members did not have available to them. Such information might come from new company research, from work done by others and reported in industry literature, from recently issued or revised RAGAGEPs, from internal and external incident investigation learnings, from the work activities discussed in the "Process Knowledge Management" element of the RBPS framework, or through development of process safety information (PSI) as defined in United States process safety regulations. New information may result in new or different conclusions or recommendations in the PHA revalidation report. Each revalidation offers an opportunity to review past risk judgments in light of current knowledge. In particular, information gleaned from incident investigations may confirm the credibility of hazards/hazardous events that seemed implausible to the prior PHA team.

Comply with Frequency Requirements. The schedule for revalidating a PHA is often dictated by a maximum allowable time interval, as discussed in Section 1.6.

Comply with Content Requirements. The need to revalidate a PHA may be driven by changes in the required content. Examples of this include the revision of the company procedure or policy for the conduct of PHAs, or applicability of new RAGAGEPs to an existing process. In the United States, state requirements may be more stringent than federal government requirements, and in the European Union, member states may impose more stringent requirements than the current revision of the Seveso Directive. The revalidation team must ensure the PHA meets **current** regulatory content requirements.

Note that an organization's understanding of its process risks can change (increase or decrease) when a PHA is revalidated. In those cases where perceived risks exceed the organization's current risk tolerance, the revalidation team may propose risk reduction measures as discussed in Section 8.3.

1.5   PHA REVALIDATION CONCEPT

The goal of a revalidation is to verify that the PHA accurately reflects the current hazards, consequences, and risk controls for the subject process and that it meets current PHA requirements (e.g., risk judgments are in accordance with current company policy and content conforms to applicable regulations).

The terms *"Update"* and *"Redo"* are used consistently throughout this book to describe two basic approaches to a revalidation:

Update. This is an incremental approach where specific changes that have been recorded and information that has been gained since the prior PHA are reviewed to produce a revalidation report that documents revisions to the prior PHA. The *Update* effort may be relatively modest for some processes (especially for established, mature processes subject to infrequent change).

Prior PHA + PHA *Update* = Revalidated PHA

For a process (including its interfacing/utility systems) having no incidents or changes, an *Update* may only be an affirmation of the continued validity of the prior PHA.

Redo. This option is usually selected when an *Update* of the prior PHA is impractical due to the nature, number, or magnitude of the (1) defects in the prior PHA, (2) incidents that have occurred since the prior PHA, (3) operating experience gained since the prior PHA, (4) changes since the original PHA (or most recent *Redo*), (5) changes in the preferred analysis methodology, or (6) changes in the PHA requirements. In this approach, the revalidation result is essentially a brand new PHA report similar to an initial PHA report. The team starts from the beginning and performs the analysis in detail, either using blank templates or using the previous PHA as a reference or guide.

PHA *Redo* = Revalidated PHA

Considerable time, effort, and thought are invested in conducting and documenting PHAs. An *Update* leverages these investments by identifying and building upon the portions of prior PHAs that are still pertinent. In most situations, the effort required to *Update* the existing PHA will be less than that required to *Redo* the PHA. However, in some circumstances, a *Redo* is more cost-effective and efficient than an *Update*. Cost considerations aside, some organizations require a *Redo* periodically to help ensure confidence in the long-term integrity and accuracy of the analysis.

An *Update* or *Redo* approach may be applied to a PHA in its entirety. However, in practice, many companies use a combination of the *Redo* and *Update* approaches when revalidating a PHA, and it is important to understand that a PHA revalidation is not always a complete *Redo* (from blank templates) or a focused *Update*. A broad spectrum of activities can be undertaken during a PHA revalidation to produce a high quality, current PHA. The *Update* and *Redo* approaches may be applied to different portions of the same PHA, resulting in a combined style of revalidation:

(Prior PHA + PHA *Update*) + PHA *Redo* = Revalidated PHA

While application of the *Update* and *Redo* approaches is discussed in detail in Chapter 5, one example of the spectrum of approaches includes the term "*Retrofit*" as it was used in the first edition of this book. A *Retrofit* is simply a *Redo* of one or more "repairable" deficiencies in the prior PHA and then an *Update* of the remaining portion of the PHA to create one correct and complete revalidated PHA.

1.6   PHA REVALIDATION CYCLE

PHA revalidations are performed periodically throughout the life cycle of a process. However, as illustrated in Figure 1-3, they are not performed in isolation. Revalidations consolidate the information gathered by other PSM elements in an operating plant and ensure the PHA provides a current assessment of the process risks.

The PHA revalidation cycle (or schedule) is the time between the previous PHA and the PHA revalidation. During this time, the facility gathers operating experience on the process. This experience can be internal (e.g., through MOC or incident investigation) or external (e.g., regulatory changes or RAGAGEP revisions).

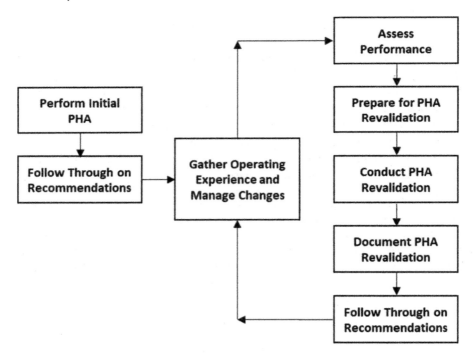

Figure 1-3 PHA Creation and Revalidation Cycle

Some facilities with mature PSM processes prefer to maintain a "living," "evergreen," or "continuous" PHA, which can be an effective way to merge the PHA and MOC processes. For these facilities, the relevant PHA documentation is revised whenever a change occurs for any reason (e.g., process technology improvement, recommendation implementation, capacity expansion, and incident response). However, such revisions to portions of the PHA are usually focused on the scope of the individual changes and **do not** revalidate the PHA as a whole. Thus, the PHA revalidation cycle is unchanged. The advantages of this practice are (1) preparation for the formal PHA revalidation can be vastly simplified and (2) it is more likely that teams reviewing changes might spot a concern related to another change because the cumulative history is documented in the PHA. Continuously revising a working copy of the most recent PHA can also shorten the PHA revalidation team meetings because the evergreen PHA should accurately document the risks associated with the current process.

A common PHA revalidation cycle is five years. This interval is specified in the United States standards/regulations of the Occupational Safety and Health Administration (OSHA) and the Environmental Protection Agency (EPA), and is the typical frequency recommended by industry associations and a frequency that has been used historically by many companies. Facilities not covered by government regulations may establish their own, appropriate revalidation frequencies. For example, some companies choose to perform PHAs on all processes, but they extend the revalidation cycle of voluntary PHAs to seven or ten years for lower hazard or non-regulatory covered processes.

When Should the PHA Revalidation Meetings Start?

What determines the required date for beginning the PHA revalidation meetings if the revalidation cycle is set at five years? Is it five years from (1) the prior PHA first meeting date, (2) the prior PHA last meeting date, (3) the date the prior PHA report was issued, (4) the date management approved the final PHA documentation, or (5) some other date? PHA meetings can span weeks or months, and the final PHA report can be issued several weeks (if not months) later.

The meetings must be started far enough ahead of the required completion date to allow a high-quality product to be produced in compliance with all requirements.

Additional guidance on establishing the PHA revalidation schedule to meet the revalidation cycle requirements is provided in Section 6.1.3.

Any facility may schedule its PHA revalidations more frequently than applicable regulations demand, either routinely or under special circumstances. Companies may choose to shorten revalidation cycles for reasons that include:

- Revalidations that are more frequent are needed to meet company loss prevention goals.

- If a major project is completed late in the revalidation cycle, it is often more effective to revalidate the PHA as part of the project hazard analysis effort rather than performing dozens of MOCs for the project and then *Updating* or *Redoing* the PHA in its entirety at a later time.

- MOCs, pre-startup safety reviews (PSSRs), and operational readiness (OR) reviews are intended to maintain the integrity of original safety features designed into the process, and to ensure that any new hazards are properly managed. However, the potential for overlooked and uncontrolled hazards exists, and the potential for such oversights increases with the number of process modifications. Some companies choose to conduct a revalidation based upon the cumulative or simultaneous number of changes (e.g., during a unit shutdown).

- Some companies have established frequencies for revalidating PHAs based upon a risk categorization (e.g., high, medium, low); the higher the risk, the more frequent the revalidation.

- Companies with multiple processes may decide to stagger the revalidation schedules for those processes to balance revalidation efforts from year to year.

- Significant incidents or an unfavorable incident trend in a process might give reason to question the adequacy of the prior PHA, prompting an expedited revalidation. Similarly, incidents at another company site, or even outside the company, may provide reasons to re-evaluate the prior PHA.

- Mergers or acquisitions may prompt management to quickly reconcile quality or protocol disparities in PHAs or to conduct new baseline PHAs based on current company standards.

- Companies might have a concern that PHAs conducted early in the development of the facility PSM program may not have been conducted rigorously or effectively, or the team may have lacked sufficient process knowledge/experience to identify and evaluate all hazards. Thus, the revalidation schedule for the facility might be accelerated.

In summary, multiple factors influence the PHA revalidation schedule for a company. The decision should be made on a process-specific basis to address needs unique to a process, facility, or company.

1.7   THE ROLE OF A PHA REVALIDATION PROCEDURE

The RBPS book identifies **maintaining a dependable practice** as a key principle to address when developing, evaluating, or improving any management system for risk [3, pp. 45-46]. Because many facilities include PHA revalidations within their HIRA program, the PHA revalidation procedure (or practice) should be clearly documented. The PHA revalidation procedure should explicitly state the benefits, value, and expectations of PHA revalidations, and lay out the requirements for performing consistent, high-quality PHA studies.

PHA revalidation procedures should:

- Establish responsibilities for revalidation roles and team members
- Establish revalidation schedules and cycles to ensure the analyses are conducted in a timely fashion
- Help users determine the appropriate methods and techniques for the study by identifying the preferred core methodology, complementary analyses, and supplemental risk assessments and by providing guidance on when and how each should be used
- Ensure the conduct and documentation of the revalidation complies with pertinent company and regulatory requirements
- Provide standard outlines of the content and format of *Redo-* and *Update*-style revalidation reports, or a desired combination thereof, so revalidations and their results are documented in a similar manner throughout a facility or company

While some organizations may need to address additional considerations (e.g., due to company-specific or regulatory requirements), a basic PHA revalidation procedure covers the activities discussed in detail in this book. Figure 1-4 shows the PHA revalidation process, including several key decision points in determining the appropriate revalidation approach or approaches. This flowchart, along with this entire book, can be used as a guide to help develop a PHA revalidation procedure.

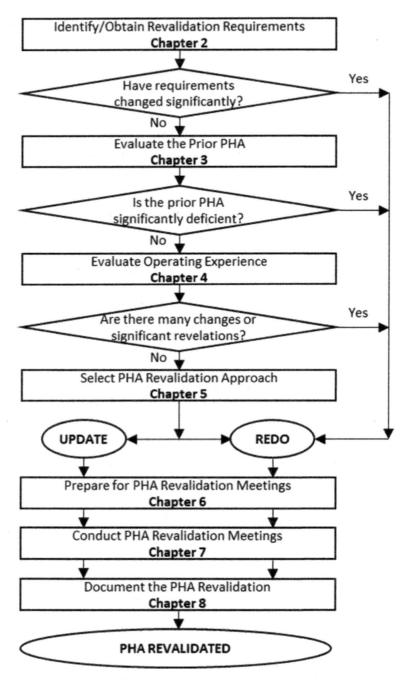

Figure 1-4 Complete Revalidation Flowchart

1.8   RELATIONSHIP OF RBPS PILLARS TO A PHA REVALIDATION

The RBPS book [3] describes the RBPS approach and recognizes that not all hazards and risks are equal. It advocates that resources should be focused on more significant hazards and higher risks. The RBPS approach is built on the four accident prevention pillars. Included is a summary of how PHA revalidation is related to each pillar and its associated elements. For detailed information on the RBPS accident prevention pillars and their elements, see the RBPS book:

Commit to Process Safety. Process safety culture is generally defined as "How we do things around here" or "How we behave when no one is watching." In a facility or company with a strong process safety culture, a PHA revalidation is an important and highly valued exercise.

In order to meet the process safety goals of an organization, a revalidation team consists of a diverse team of well-trained process experts who understand the importance and value of the revalidation.

Conversely, in a facility where process safety culture could be improved, the PHA revalidation exercise is seen as a rote task that must be completed (e.g., to satisfy regulatory or company require-

> **Process Safety Culture**
>
> In a facility with a strong process safety culture, there is trust that the PHA revalidation will be of value and the results of the study will be acted upon. This trust and outreach/communication flows from operations, through the PHA revalidation team, to management, and then back again.

ments). The goal for such a PHA is to be able to claim it was done on time but required little or no effort to perform and created minimal additional work for follow-up.

Understand Hazards and Risk. Process knowledge is a critical key to the success of any PHA revalidation and includes both process documentation and the competency of process experts (with many areas of experience including operations, engineering, and application of PHA methods). A revalidated PHA that incorporates current knowledge of the process and its hazards is foundational for all the other RBPS elements that rely on the PHA, such as the asset integrity of safety-critical equipment, MOC, or emergency response planning.

HIRA is a term that encompasses all activities involved in identifying hazards and evaluating risk at facilities, throughout their life cycle, to make certain that risks to employees, the public, or the environment are consistently controlled within the organization's risk tolerance. A PHA is one form of HIRA and is sometimes required to meet specific regulatory requirements. The concepts of revalidation described in this book can be applied to a PHA or HIRA to address the key principles of RBPS:

> **The PHA can be used to:**
>
> - Identify which safeguards should be classified as critical
> - Help develop preventive maintenance schedules and priorities for safety-critical instrumentation
> - Identify scenarios for emergency response training drills
> - Identify which written operating procedures should be classified as critical
> - Help develop a training program for new operators
> - Establish the baseline for MOC reviews

- Maintain a dependable practice
- Identify hazards and evaluate risks
- Assess risks and make risk-based decisions
- Follow through on assessment results

Manage Risk and Learn from Experience. From the time an initial PHA is completed on a process until the PHA is revalidated, risk is being managed throughout the revalidation cycle. Changes are implemented, incidents are investigated, work is performed, operating experience is gained, and external conditions may change. The PHA revalidation is an opportunity for the facility to look back at this operating experience and learn from it to reduce risk in the process. Whether the PHA revalidation is a focused *Update*, a complete *Redo*, or a combination of these approaches largely depends on what has happened in the process since the last PHA.

2 PHA REVALIDATION REQUIREMENTS

E. I. DuPont, founder of E. I. DuPont de Nemours and Company in 1802, recognized the inherent risks in the manufacture of gunpowder. From the beginning, DuPont required hazard studies of their processes with a strong emphasis on facility siting. Recommendations from those studies were very effective in reducing the risks of injuring workers or destroying company assets [22]. In the 1950s, Imperial Chemical Industries (ICI) invented the Hazard and Operability (HAZOP) technique, and as its safety and economic benefits were demonstrated, ICI began requiring that all its chemical processes be subjected to a series of hazard studies [23] [24]. ICI's Hazard Study 3 would be called a process hazard analysis (PHA) today, and ICI's Hazard Study 6 is the initial revalidated PHA.

However, not every company voluntarily embraced PHAs. Major accidents at chemical plants in Flixborough (England, 1974), Seveso (Italy, 1976), and Bhopal (India, 1984) shocked surrounding communities and the world. With each additional accident, public pressure escalated on officials to reduce the risk of hazardous chemicals injuring workers, the public, and the environment. The result has been a steadily increasing patchwork of laws and regulations intended to improve chemical process safety around the world. Most of these regulations require some sort of PHA that must be periodically revalidated.

Thus, two sources guide PHA and PHA revalidation requirements. External requirements are the legally enforceable requirements that a facility must meet to operate within a specific jurisdiction. Internal requirements are those that a company imposes on itself to achieve its organizational risk goals.

2.1   EXTERNAL LEGAL/REGULATORY REQUIREMENTS

The historic pattern is that government authorities respond to major accidents by enacting laws and promulgating regulations that give them the power to proactively establish, audit, and enforce new process safety requirements. For example, a runaway chemical reaction in July 1976 led to a dioxin release that affected thousands of people near Seveso, Italy [25, pp. 295-299]. That prompted a series of government responses, and the Council of the European Union issued

Directive 82/501/EEC in June 1982. This Directive and subsequent revisions, commonly referred to as Seveso I, II, and III, require the member states of the European Union "to take all measures necessary to prevent major accidents and to limit their consequences for man and the environment." Establishment operators must prepare a safety report and submit it to the competent authorities of the Member State where the facility is located. The safety report must be reviewed and updated as necessary [revalidated] at least every five years, but the Directive also triggers revalidation "at the initiative of the operator or the request of the competent authority, where justified by new facts or to take account of new technical knowledge about safety matters."

In their implementation of the Directive, European Union member states also have the latitude to impose additional or more restrictive requirements. For example, the Netherlands requires an External Safety Report that includes a QRA not required by the Seveso Directive. Authorities in the Netherlands chose a quantitative approach because of their belief that qualitative opinions would not provide an adequate basis for making decisions about safety.

United States regulations followed a similar arc. The 1984 methyl isocyanate release that led to thousands of fatalities in Bhopal, India [25, pp. 25-30], triggered a chain of events that culminated in the United States Clean Air Act Amendments of 1990. Among those amendments were requirements that both OSHA and EPA develop regulations intended to prevent accidental releases of highly hazardous materials. In May 1992, OSHA issued its PSM Standard. The PSM Standard applies to those processes (i.e., "covered processes") containing more than a specified threshold quantity of a defined highly hazardous material (i.e., "covered chemical"), and it also applies to the manufacture of explosives.

Among the provisions of the PSM Standard is a requirement to conduct PHAs for covered processes. In the United States, OSHA provides general requirements for the conduct of the PHA, addressing issues such as scheduling, team make-up, permissible analysis techniques, and topics that must be addressed. In addition to the core analysis of process hazards, risk controls, and the consequences of safeguard failures, the regulation also requires special attention to facility siting and human factors issues. The requirements also include the subject of this book – that the PHA be revalidated – at an interval not to exceed five years. The OSHA PHA and revalidation requirements can be seen in their entirety by searching for "1910.119" on OSHA's website.

EPA similarly issued its Risk Management Plan (RMP) Rule in June 1996. Unlike OSHA's uniform requirements for PSM programs to protect workers, EPA established three different RMP program levels to address facilities posing different levels of risk to the community. The accident prevention requirements for EPA's most rigorous level, Program Level 3, generally parallel those of the OSHA PSM Standard (although EPA's lists of "regulated substances" [i.e., covered chemicals] and their threshold quantities are somewhat different), including the requirements for conducting PHAs and subsequently revalidating them every five years. For its less rigorous Program Level 2, EPA requires a hazard review, which is a less detailed version of a PHA, but it must still be revalidated at least every five years. The EPA PHA, hazard review, and revalidation requirements can be seen in their entirety by searching for "40 CFR 68" on EPA's website.

While the OSHA PSM and the EPA RMP Program Level 3 requirements are very similar, they have an important difference in regulatory focus. OSHA has responsibility solely for the safety of facility workers, while EPA has responsibility for the safety of off-site populations and sensitive environments. These different regulatory responsibilities are reflected in the specific PHA requirements.

Beyond the federal requirements discussed in this chapter, individual states in the United States may impose additional or more stringent requirements, and some of them affect PHA revalidations. For example, United States federal regulations only require qualitative consequence analyses in PHAs. In New Jersey, the PHA must include "Table(s) of the process hazard analysis results giving the release point and corresponding release scenario of the potential basic (initiating) and intermediate event sequences, the corresponding estimated quantity or rate and duration of releases ..." [26]. California requires that PHAs for oil refineries include a damage mechanism review that verifies the materials of construction are appropriate for their application and are resistant to potential damage mechanisms [27].

Legal requirements for chemical process safety have proliferated around the world, wherever highly hazardous materials are processed. Australia, Brazil, China, India, Mexico, Russia, and South Korea are among the growing number of countries that have specific process safety regulations. Although the terminology and technical details vary, they commonly require development of a baseline PHA and periodic revalidation of it, just like the United States and European Union.

2.1.1   General Obligations

Beyond explicit legal obligations, companies certainly have an ethical obligation to protect workers, the public, and the environment. Another concept is that individuals or companies should exercise reasonable care while performing any acts that could foreseeably harm others. For example, the Standards Council of Canada issued its *Process safety management* standard (CAN/CSA-Z767-17) in

August 2017. Compliance is a general obligation, so this standard has no specific list of chemicals or quantities. The Canadian PSM standard requires an initial risk assessment (PHA) that must be revalidated at least every five years.

It is sometimes argued that specific laws or regulations in one jurisdiction can create obligations elsewhere. This was particularly evident in the prosecutions following the 1998 explosion at the Longford gas plant in Australia [25, pp. 167-174]. It was argued that the site operator's performance of retrospective HAZOPs (as regulations required elsewhere) created an obligation for them to be performed for facilities in Australia, even though there were no specific Australian PHA requirements at the time. (Australia enacted specific regulations on Major Hazard Facilities in 2000). Therefore, organizations should strongly consider being globally consistent in how they perform PHAs and revalidations.

2.1.2   Specific RAGAGEPs

With respect to process safety, companies around the world usually rely on their compliance with RAGAGEPs to show that they have fulfilled their general obligations. For example, the adequacy of overpressure protection is one of the most important factors in risk judgments faced by any PHA team. Compliance with the American Society of Mechanical Engineers (ASME) Boiler and Pressure Vessel Code (or an international equivalent, such as the European Union Pressure Equipment Directive) is typically considered a very reliable safeguard against vessel rupture due to overpressure. Examples of related issues a revalidation team should consider are whether:

- Process changes have been appropriately managed such that no new scenarios might exceed the design basis for the pressure relief devices.
- Reductions in maintenance staffing or inspection/test frequencies for the pressure relieving devices or protected equipment have affected code compliance.
- Mechanical integrity of the protected equipment has been maintained within acceptable limits for its pressure rating.

Boiler and pressure vessel codes are uniquely important in that many jurisdictions require compliance by law; however, organizations typically have more flexibility in their approach to compliance with other RAGAGEPs.

Multiple RAGAGEPs might apply to a process being evaluated in a PHA, so it would be unreasonable to expect that any initial PHA team or revalidation team would be aware of every possible RAGAGEP. Hence, as mentioned in Chapter 1, organizations (e.g., companies, trade associations, and professional societies) often develop checklists to remind people of key points to verify during original designs, PHAs, and revalidations. An excellent example of this is Bulletin 109, published by the International Institute of Ammonia Refrigeration for the mutual benefit of its membership [28]. It compiles safety criteria for a variety of equipment (e.g., compressors, condensers, evaporators, and piping) normally used in refrigeration systems. The following paragraphs briefly describe some common RAGAGEPs that PHA revalidation teams might consider:

International Electrotechnical Commission (IEC) 61508. "Standard for Functional Safety of Electrical/Electronic/Programmable Electronic Safety-Related Systems" [29]. This standard covers safety-related systems, which may include anything from simple actuated valves, relays, and switches up to complex programmable logic controllers. The standard specifically addresses safety instrumented systems and how their failure to execute their safety functions either on demand or continuously can result in safety-related consequences. The overall program to ensure system reliability is defined as "functional safety." IEC 61511 is the process industry derivative of IEC 61508. Risk judgments in a PHA of an automated process may involve supplemental risk assessments (e.g., LOPA) that are contingent upon the reliability of safety systems covered by this standard; in these cases, it is essential to have a revalidation team member(s) familiar with its guidance.

American National Standards Institute/National Fire Protection Association (ANSI/NFPA) 70. "National Electric Code" [30]. As its title indicates, this code applies to a wide range of electrical power systems and their components. Various chapters address everything from basic definitions and rules for installations (e.g., voltages, connections, markings, and circuit protection) to general-purpose equipment (e.g., conductors, cables, receptacles, and switches) to special equipment and conditions (e.g., signs, alarms, emergency systems, and communications). Articles 500 to 506 are of particular interest to revalidation teams trying to ensure that process changes have not created or altered areas with potentially explosive concentrations of dusts or vapors. European Directive 94/9/EC, commonly referred to as the Atmospheres Explosible (ATEX) Directive, similarly addresses equipment in potentially explosive atmospheres.

NFPA 652. "Standard on the Fundamentals of Combustible Dust" [15]. Dust fires and explosions have resulted in fatalities and extensive property damage. Yet the materials involved may be as seemingly innocuous as sugar or polyethylene, so they are not classified as highly hazardous materials. Dust hazard analyses are recommended by the general NFPA 652 standard, along with several material-specific standards:

- **NFPA 61.** "Standard for the Prevention of Fires and Dust Explosions in Agricultural and Food Processing Facilities"
- **NFPA 484.** "Standard for Combustible Metals"
- **NFPA 654.** "Standard for the Prevention of Fire and Dust Explosions from the Manufacturing, Processing, and Handling of Combustible Particulate Solids"
- **NFPA 655.** "Standard for Prevention of Sulfur Fires and Explosions"
- **NFPA 664.** "Standard for the Prevention of Fires and Explosions in Wood Processing and Woodworking Facilities"

If dust hazards may be present in a process, dust hazard analyses based on such standards are often performed as complementary analyses in the PHA and are subject to revalidation.

API-RP-752, "Management of Hazards Associated with Location of Process Plant Permanent Buildings" [5], **API-RP-753,** "Management of Hazards Associated with Location of Process Plant Portable Buildings" [6], and **API-RP-756,** "Management of Hazards Associated with Location of Process Plant Tents [7]. Most PHAs require special consideration of facility siting issues. Many companies use the guidance in these RAGAGEPs, and to perform analyses of buildings intended for occupancy, and the results of such studies are considered when addressing facility siting in related PHA(s) and making related risk judgments. In these cases, a revalidation team should have such study results available and consider whether (1) any changes to the process affect the potential for vapor cloud explosions, fires, and toxic gas release scenarios or (2) the use or occupancy of buildings near the process have been adequately managed with respect to such study results.

2.2   INTERNAL COMPANY POLICY REQUIREMENTS

Many companies that perform PHAs have written company policies governing PHAs and related activities such as revalidations. Such internal PHA policies can be broadly categorized as compliance-driven; environmental, health, and safety (EHS)-driven; or value-driven policies, and they often correlate with the maturity of the organization's process safety culture.

2.2.1 Compliance-Driven Policies

Companies with compliance-driven policies typically just mirror the minimum regulatory requirements of the jurisdiction(s) where they have operating facilities. In those policies, the definition of PHA revalidation is relatively straightforward because the organization simply wants confirmation that current regulatory requirements have been met. For example, if regulatory requirements do not explicitly require a *Redo* of PHAs at some frequency, neither would a compliance-driven policy. A company with a compliance-driven policy must ensure that revalidation teams understand that compliance requires more than a perfunctory review. The team must critically evaluate the prior PHA and revise it as necessary to accurately reflect the current process safety risks.

In many jurisdictions, the regulatory requirements do not explicitly define tolerable risk, so each organization must define its own. For example, an organization might adopt "compliance with RAGAGEPs" as their definition of tolerable risk. In such companies, if a pressure vessel had a properly designed pressure relief valve and the PHA team identified no scenarios that could exceed its capabilities, the team might judge the risk of vessel rupture due to overpressure tolerable with no further risk reduction warranted. A more challenging situation arises when tolerable risk is defined to be ALARP. The fluid notion of tolerable risk can change with time and circumstances; so, the revalidation team may judge risks differently and identify new ways to reduce risks deemed tolerable or ALARP in the prior PHA.

2.2.2 EHS-Driven Policies

Companies with EHS-driven policies typically expand the list of processes subject to PHA. For example, when a member of the American Chemistry Council commits to the Responsible Care® initiative, they indicate that they will follow Responsible Care Guiding Principles that include (1) designing and operating facilities in a safe, secure, and environmentally sound manner and (2) instilling a culture of continually identifying, reducing, and managing process safety risks. Thus, they often require PHAs for all their operating assets that pose process safety risks.

Companies with EHS-driven policies may also expand the content of each PHA. Their policies usually incorporate the best features of relevant regulatory requirements, so they have a globally consistent policy. Therefore, for example, a company headquartered in the European Union might require all PHAs to include a safety case to document how it plans to operate a facility safely and effectively manage its risks. EHS-driven policies often include the use of supplemental risk assessment techniques, such as LOPA or bow tie analysis, to help ensure consistent risk judgments and resource allocations. To support such analyses, the company must develop some form of risk matrix or risk tolerance criteria to guide PHA revalidation teams when revalidating supplemental risk assessments.

These policies directly affect the revalidation efforts. Inclusion of all operating assets will increase both the number of PHAs and the number of analysis nodes therein. The inclusion of supplemental risk assessments will increase the time required to evaluate any added or modified loss scenarios. Similarly, expanding the PHA to include complementary analyses (e.g., dust hazard analyses, reviews of inherently safer technologies, structural analyses of potentially occupied buildings, or checklists verifying compliance with substance-specific or technology-specific guidelines), will expand the scope and complexity of revalidations.

2.2.3   *Value-Driven Policies*

Companies with value-driven policies try to extract maximum benefits from resource expenditures. To achieve their goal of safe, sustainable production, they typically expand the scope of their PHAs beyond EHS risks to include economic risks such as poor quality, low yields, and unplanned downtime. Long before there were any regulatory requirements, some companies voluntarily developed and used the methods now associated with PHAs for the demonstrable economic, safety, and environmental benefits. (Modern HAZOPs were originally called Operability and Hazard Studies.) Value-driven policies do not necessarily increase the number of the PHAs. Sometimes, the intent is simply to obtain the maximum value from PHAs beyond the mere compliance that regulations demand. However, value-driven policies increase the size and complexity of PHAs (and thus revalidation efforts) by including many scenarios (consequences and their related causes and controls) with economic-only impacts that are not evaluated or documented for compliance-driven or EHS-driven policies. In addition, value-driven policies may expand the scope of MOC and incident investigation programs, which will increase the number of documented changes and incidents to be reviewed in an *Update*-style revalidation. Value-driven policies must be carefully managed to ensure that the primary goal of improving process safety is not subordinated to other considerations.

2.3   INTERNAL COMPANY DRIVERS THAT IMPACT REVALIDATION

All aspects of a PHA (its core analysis, complementary analyses, and any supplemental risk assessment) must be revalidated on the schedule prescribed by regulation or internal policy, as previously discussed. However, situations do arise that may warrant revalidation of all or a portion of a PHA on an accelerated schedule. For example, the PHA for an acquired asset will have to be revalidated in accordance with the purchaser's policy, using its preferred method and risk tolerance criteria. The question that most often arises is whether to revalidate the acquired asset PHA promptly after acquisition or on the schedule established by the previous owner. Considering the uncertainties (e.g., staffing, operational, and maintenance changes) associated with most acquisitions, accelerating the revalidation schedule is not typically beneficial. Exceptions are situations where pre-acquisition due-diligence reviews identified significant PHA deficiencies or where the previous owner's risk tolerance was significantly different from that of the purchaser.

Mergers may also trigger revalidations, but the situation is much more complex than an acquisition. In a merger, it is likely that a new PHA policy will be synthesized from the existing policies of the parties involved. Once issued, that new policy would then affect all upcoming revalidations. In most merger situations, PHA revalidations continue on their original schedule.

A major unit expansion, revamp, or repurposing may also warrant an accelerated revalidation. For example, a debottlenecking project might collectively involve dozens of physical changes and alter the flow velocities through most of the existing equipment. The normal MOC process could be followed, and the revalidation could occur on its normal schedule. However, if the existing PHA is already three or four years old, and the proposed change is going to add several new equipment items and/or significantly affect most of the existing equipment, the MOC will require about as much effort as a full revalidation. Therefore, rather than perform a large and complicated MOC, followed soon thereafter by a large and complicated revalidation, it is usually much more efficient to combine the two activities. With that approach, at the conclusion of the change, the company will have a fully revalidated PHA, and the next revalidation cycle will start.

Another reason to accelerate revalidation is the acquisition of significant new process knowledge or operating experience indicative of weaknesses in the current PHA. Laboratory testing may have discovered a new mechanism for runaway reactions. Inspections may have discovered unexpected corrosion due to contaminants in raw materials from certain suppliers. A serious incident may have revealed deficiencies in the information available for operators to diagnose process upsets correctly. Any such information could justify revalidation of the current PHA to ensure it considers known hazards and accurately assesses process risks.

Sometimes, internal requirements may trigger a revalidation simply because there have been a large number of changes. Each change brings the potential for overlooked hazards or compromised risk controls, and the potential for such oversights increases with the number of process modifications. The potential for such oversights also increases when multiple changes are made simultaneously. For example, some changes can only be made during a unit shutdown, and to minimize downtime, multiple teams usually implement multiple projects in parallel. Thus, some companies choose to conduct a revalidation when the number of cumulative or simultaneous changes exceeds their threshold.

Examples of Changing Company Requirements Affecting the Revalidation

Example 1 – Company A acquires a processing unit that is subject to regulatory PHA requirements. The prior owner used the What-If/Checklist technique for the core PHA and met all other regulatory requirements. Company A's preferred core analysis method is HAZOP. The question is whether the PHA should be *Redone* as soon as possible or used "as is" until the required regulatory revalidation date three years hence. Because there is no indication that the risk judgments would be substantively different as a result of changing analysis techniques, Company A decides its resources would be better used elsewhere and waits to revalidate the PHA.

Example 2 – Company B acquires a plant that uses natural gas extensively in its metallurgical processes. However, the amount of flammable material is well below regulatory thresholds. The prior owner had voluntarily done a HAZOP of the existing furnace during its initial design, but that study had never been revalidated. Soon after the acquisition, Company B decides to *Update* the HAZOP of the furnace and perform complementary analyses of human factors and facility siting to conform with its corporate requirements of subjecting all its hazardous processes to PHA.

The desire to smooth the annual revalidation workload is another reason to accelerate a revalidation. Often when organizations initiate a PHA program, they prioritize their highest hazard, most complex units. In addition, they often mandate that the PHAs be done as soon as possible. This creates a spike in the PHA workload during the first few years of the PHA cycle, and then a relative lull.

When those initial PHAs must be revalidated, there is an echo effect, with similar spikes and lulls in the revalidation workload. Revalidating some of the PHAs before their due date is an effective way to average the workload year-to-year and simplify management of the entire program.

Internal drivers may also be related to a company's desire to keep risk as low as reasonably practicable in situations where regulatory requirements are not explicit. For example, the United States EPA RMP Rule requires a facility to update and resubmit its RMP if it begins to use more than a threshold quantity of a regulated substance that it has not previously reported for the process in which it is being used. The facility must update and resubmit its RMP to reflect the change by the time the regulated substance exceeds the threshold quantity in the process. In that case, it would be prudent to revalidate the PHA before submitting the revised RMP, even though a PHA revalidation is not explicitly required prior to resubmission.

An updated RMP is also due within six months of a change that requires a revised offsite consequence analysis. In that case, a company might choose to *Update* the relevant PHA(s), so the related scenario(s) and facility siting information is consistent between the PHA, the revised off-site consequence analysis, and the updated RMP submission. Similarly, if a facility experiences an accidental release that requires updating the five-year accident history section of the RMP submittal, the organization may also want to *Update* the relevant PHA(s) to show consideration of that incident, its actual and potential consequence(s), and any new controls added to help manage the risk.

2.4   PRINCIPLES FOR SUCCESSFUL DEFINITION OF REVALIDATION REQUIREMENTS

Actual experience in conducting revalidations by PHA practitioners across a number of companies has highlighted some keys for success, as well as some things that can impede success. While none of the items listed are absolute rules, they do provide valuable guidance.

Successful Practices:

- Reviewing the current legal requirements applicable to the PHA to see if the prior PHA is compliant. If not, planning to correct the non-compliant portions during the next revalidation (*Update* or *Redo*)
- Reviewing the current corporate requirements applicable to the PHA to see if the prior PHA is compliant. If not, planning to correct the non-compliant portions during the next revalidation

- Revalidating the PHA to ensure current company risk practices and procedures are being followed (If an asset has been acquired by a new company). This revalidation may be performed on an accelerated schedule
- Confirming management agreement that the revalidation will use the same consequences of interest as the prior PHA or revising as necessary
- Looking for opportunities to maximize the value of PHA revalidation activities beyond minimal regulatory requirements
- Ensuring outside personnel being used for PHA facilitation or as subject matter experts (SMEs) are knowledgeable in all requirements prior to initiating the revalidation sessions

Obstacles to Success:

- Trying to simply *Update* a PHA when fundamental regulatory or organizational requirements have changed
- Trying to only revalidate a portion of the process when requirements have changed
- Trying to revalidate a PHA when organizational requirements are in flux due to a merger or acquisition; prior to initiating a revalidation, the requirements should be defined
- Neglecting to identify and explain requirements to the revalidation team (including contractors and SMEs) prior to the meeting

3 EVALUATING THE PRIOR PHA

The primary objective of a PHA revalidation is to ensure the PHA accurately describes and evaluates current risks. To accomplish that, the team should consider the current hazards, current causes, current consequences, and current risk controls. The purpose of evaluating the prior PHA is to assess how closely that analysis matches the current reality, which is the key determinant of the best revalidation approach. Presuming regulations or policies do not dictate the use of the *Redo* approach as discussed in Chapter 2, any decision to set aside previous efforts and completely *Redo* the prior PHA with a clean sheet of paper and a new team, risks losing all the effort and experience that were embodied therein. For example, technology-specific hazards that had been learned by the operating company might have prompted some of the loss scenarios documented in the prior PHA. That potential loss must be balanced against the possibility of relying too much on outdated information that is not relevant to the unit's current risk profile along with any errors or omissions of previous teams.

Thus, given the spectrum of revalidation approaches ranging from a complete *Redo* to a focused *Update*, several factors should be considered when selecting an optimum approach. The dominant considerations generally fall into the following categories:

- Changes in PHA requirements
- Methodology used for the prior PHA
- Inputs for the prior PHA
- Overall quality and completeness of the prior PHA
- Changes made to the process since the prior PHA
- Consistency of PHA risk judgments with operating experience

As illustrated in Figure 3-1, changes in PHA requirements were addressed in Chapter 2. This chapter focuses on evaluating the acceptability of the prior PHA. Evaluating what has transpired since the last PHA and assessing the consistency of risk judgments with operating experience (including learnings from incidents) are addressed in Chapter 4.

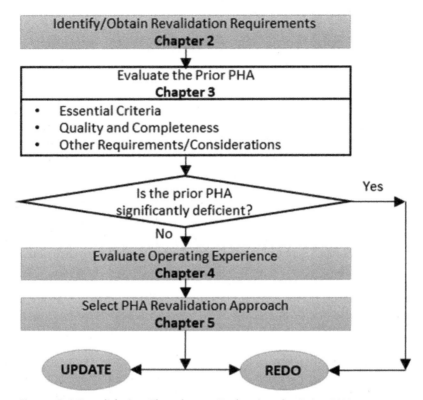

Figure 3-1 Revalidation Flowchart – Evaluating the Prior PHA

The *Update* approach is generally followed when it is expected to be less time-consuming and more cost-effective. A general assumption underlying any mention of the *Update* approach in this book is that the prior PHA was conducted and documented in an acceptable manner and identified the known and foreseeable hazards of the process, the related risk controls (safeguards) employed to manage risk to a tolerable level, and the consequences of failure of those controls. Therefore, a key step in planning a PHA revalidation is to determine whether the prior PHA is a suitable foundation on which to build the revalidation effort by answering the question **"Is the prior PHA significantly deficient or unacceptable?"** If the answer to this question is "Yes," this is reason enough to *Redo* the PHA.

In order to judge acceptability of the prior PHA, one should consider:

- Does it meet essential criteria? (See Section 3.1.)
- Is it of good quality and complete? (See Section 3.2.)
- Are there other reasons to *Redo* the prior PHA? (See Section 3.3.)

Appendix A includes a sample checklist that can be used to determine if the essential criteria are satisfied, and Appendix B includes a checklist that can be used to evaluate quality and completeness of the prior PHA. If significant deficiencies are found, management should be advised of the issues so they can be addressed outside of the revalidation process. The steps for evaluating the prior PHA are depicted in Figure 3-2.

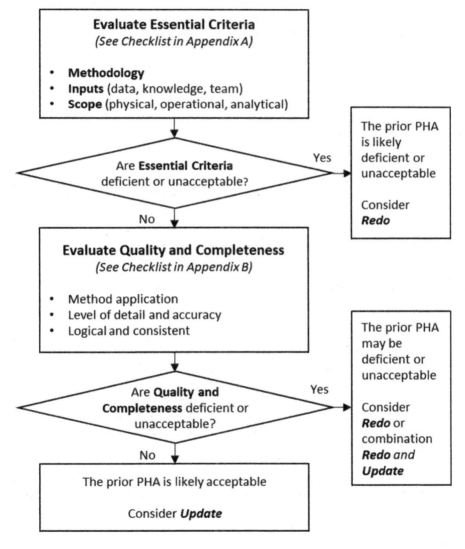

Figure 3-2 Steps for Evaluating the Prior PHA

3.1   PRIOR PHA ESSENTIAL CRITERIA

3.1.1   Prior PHA Methodology Used

The prior PHA methodology refers to its core methodology and any complementary analyses as discussed in Section 1.2. Aside from policy or regulatory demands, two key questions should be considered when determining whether the prior PHA methodology was appropriate:

1. Was the PHA methodology appropriate for the complexity of the process?
2. Did the PHA methodology comprehensively identify the hazards of the process, the engineered and administrative risk controls, and the worst credible consequences assuming failure of all those controls?

For example, PHA results comprised of only a short one-page checklist with yes/no answers and related comments would be insufficient as a core methodology for a PHA of a complex and hazardous unit involving reactions, separations, and so forth. Likewise, an unstructured What-If Analysis only identifying loss scenarios where the team had a concern or recommendation would generally be considered inadequate as a core PHA methodology. However, a thorough What-If Analysis combined with equipment- or process-specific checklist analyses might be an appropriate PHA methodology for some processes.

Under most circumstances, a PHA including a properly applied and fully documented HAZOP as the core methodology will identify the process hazards, risk controls, and consequences of failure of the controls for each section of a process. A PHA structured with a well-organized and fully documented What-If/Checklist Analysis using appropriate process- and equipment-specific checklists can also accomplish these objectives. For processes involving compressors, centrifuges, or highly automated systems, an FMEA can be conducted and documented in a manner that meets these objectives.

Any of these core methodologies can, by themselves, be applied and documented in a manner that identifies process hazards, risk controls, and consequences of failure of the controls. However, the most comprehensive PHAs also include complementary analyses, and these are often in the form of checklists.

Checklists used as complementary analyses in a PHA tend to take one of three forms:

1. **Process-specific checklist.** For some high hazard chemicals (e.g., anhydrous ammonia, chlorine, ethylene oxide, liquefied petroleum gas [LPG]), industry organizations have developed standards with specific requirements or recommendations for safe design, operation, and maintenance of processes involving those chemicals. Such standards can be used as checklists in a PHA. Additionally, process-specific operating and/or maintenance procedures are sometimes used as "checklists" to help identify activities where human error may become an issue. Any document that contains relevant information about the process and its operation can be the basis for a process-specific checklist.

2. **Equipment-specific checklist.** For equipment such as pressure vessels, piping, and electrical components, written standards are considered to be recognized and generally accepted good engineering practices (RAGAGEPs), and these can be used as checklists in a PHA. Some RAGAGEPs may only be mentioned in a PHA as having been considered by the PHA team and not documented explicitly in checklist form. Where a Checklist Analysis based on RAGAGEPs is

    > **Example – RAGAGEP Checklist Question**
    >
    > When evaluating combustible dust hazards in grain elevators, a checklist based on NFPA 61 would likely ask if conveyor belts and lagging are fire resistant, oil resistant, and resistant to electrical charge accumulation (surface resistivity less than 300 mega-ohms per square unit of area).

    documented in a PHA, it should include a reference to the related standard(s), including their edition or effective date. If specific RAGAGEPs have been revised since the prior PHA in a manner that affects the related checklists, the affected checklist questions and answers should be revisited during the revalidation to ensure accurate responses.

3. **General checklist.** General checklists are collections of questions that organizations find useful for ensuring completeness and consistency in identifying hazards and risk controls. In PHAs, general checklists are usually organized by functional area or topic, such as facility siting (See Appendix E.), human factors (See Appendix F.), and external events (See Appendix G.).

Many of these checklists are very detailed and may overlap with topics discussed in a HAZOP, What-If/Checklist, or FMEA core analysis. Where overlaps occur, the revalidation team will need to *Update* the checklist answers along with the core analysis (and cross-referenced, where practicable) to maintain consistency throughout the PHA. If the review of these checklists finds gaps, such as required topics that are not included or questions that were not answered, the revalidation team will need to correct those deficiencies. See Section 3.2.2 for additional discussion about evaluating checklists in the prior PHA.

3.1.2   *Prior PHA Inputs*

Inputs to the prior PHA should have included (1) accurate and complete information such as safety data sheets (SDSs), piping and instrumentation diagrams (P&IDs), and relief valve calculations, and (2) a good understanding of the process and how it was intended to be operated and maintained. Factors affecting the quality of these inputs include the diversity of the PHA team's skills (e.g., operations, engineering, instrumentation and controls, maintenance), the depth of the team's knowledge/experience regarding the process, and the accuracy and completeness of process safety information available to the team. The extent to which such inputs were provided, and the quality of those inputs, affects the acceptability of the prior PHA. The following paragraphs describe the expectations for prior PHA inputs and methods to discern their quality.

Data and Information. In the CCPS book *Guidelines for Risk Based Process Safety* [3, p. 170]; the term *process knowledge* broadly refers to any information that can easily be recorded in documents, such as (1) written technical documents and specifications; (2) engineering drawings and calculations; (3) specifications for design, fabrication, and installation of process equipment; and (4) other written documents such as SDSs. United States regulations use the term "process safety information" to refer to a specific data package describing chemical hazards, process technology, and process equipment. Documentation of current process knowledge should be readily available to, and frequently used by, PHA teams. However, the PHA team may need to refer to other information, such as:

- Operating, maintenance, and emergency procedures
- Control and safety system functional specifications
- Facility plot plans
- Relevant incident investigation reports
- Other complementary analyses or supplemental risk assessments such as facility siting studies, LOPA studies, or Human Reliability Analyses

- Mechanical integrity inspection, test, and preventive maintenance plans and programs
- Manuals and other data provided by equipment manufacturers and technology integrators
- Corporate or facility engineering specifications
- Design philosophy documents

Not all the listed documentation is required for every PHA, and other technical information may be needed to support a particular PHA. For processes covered by United States PSM regulations, the PSI is the legally required set of data that must be available to the PHA team.

Evaluating whether the prior PHA team had access to, and used, adequate process knowledge documentation requires some investigation. Reviewing the prior PHA report should reveal the primary information referenced, such as specific revisions of P&IDs and other drawings. A sampling of the revision histories and dates can be examined to see if the revisions were current at the time of the prior PHA. It may also be possible to ask some members of the prior PHA team what information they recall using in the PHA. For example, if they often took credit for operator response in the field as a safeguard, ask whether team members recall reviewing a plot plan or other layout information to confirm distances, spacing, and other factors affecting an operator's ability to respond successfully. Negative or inconsistent responses from prior PHA team members may indicate a deficiency in the data and information inputs to the prior PHA.

> **Efficiency Tip**
>
> Developing a short checklist of questions for prior PHA team members will help gather needed information consistently and efficiently.

Reviewing the prior PHA recommendations and results may also reveal information deficiencies. For example, a recommendation that reads "Determine if the design basis for the relief system for the low-pressure separator includes loss of level control in the high-pressure separator and take appropriate action..." implies either (1) the team may not have had access to all relief valve sizing documentation or (2) the sizing documentation may have been incomplete, non-specific, or difficult to understand. Similarly, a recommendation that reads "Field verify the P&IDs" almost certainly means the team seriously questioned the accuracy of the process drawings. If a significant fraction of the team's recommendations includes a request to do an engineering evaluation or gather more information, one may assume the prior PHA team was lacking some of the important process knowledge documentation.

Process Understanding and Team Make-Up. Evaluating whether the prior PHA team included all the required and needed personnel involves understanding the guidance in Chapter 2 (requirements based on regulations and company policy) and the discussion in this section on prior PHA input. However, it is possible to have all the "right" personnel in the room and still not have sufficient understanding of process operations. Operations personnel on the team need to have an appreciation for how the process works. Simply knowing which valves to turn, when to turn them, and how far to turn them is only part of what is needed. Similarly, an engineering degree alone is not equivalent to engineering expertise. Engineering personnel on the team need to have a deep understanding of the chemical and physical processes (i.e., the chemistry, material and energy balances, and equipment design intentions).

While it may be difficult to evaluate whether the prior PHA team had all the required process knowledge documentation, it is even more difficult to determine whether team members had the needed process understanding. Clues are sometimes evident in the PHA documentation. For example, listed causes that cannot logically produce the claimed consequences suggest that team members lacked basic understanding of the process or loss scenarios. It may also indicate pre-conceived ideas that were not accurate. It may be possible to discuss the prior PHA with team members who participated and management personnel who were involved in assigning the team members. The answer might be something like "We really couldn't assign any of the senior process engineers for the unit to that PHA team because they were assigned full time to support the team designing Unit II," or it might be "It was absolutely our best team for that unit."

Apart from company policy requirements, there are no absolute criteria for assessing the adequacy of knowledge and experience on the prior PHA team. However, two important considerations include:

1. **Experience.** No matter how talented, no one in the first year of their assignment will have the same level of process understanding they will have a year later, and the year after that. Thus, some organizations establish minimum experience requirements for PHA team members, and if those requirements were waived for multiple team members or required team members in the prior PHA, it may be appropriate to *Redo* the PHA.

2. **Related knowledge.** For example, if an engineer spent two years as the process control engineer for a process and then moved into the process engineer role, they would have a significant head start on learning the process compared to someone hired without that background. Likewise, someone who served as a shift leader

before becoming process engineer would understand the process from a broader perspective.

Finally, experience need not be limited to years in a particular job. Consider the extreme case of a team that conducts an initial PHA of a completely new process. Since the process has not yet started up, no one on the team has any actual operating experience with the specific process at the proposed production scale. While staffing of initial PHA teams for new processes is not a PHA revalidation issue (the initial PHA organizers needed to address that issue), the team that first revalidates the PHA often has significantly more knowledge of and experience with the actual process and its operations than the initial PHA team. Therefore, for the first revalidation, "lessons learned" and operating experience gained since initial startup should be considered along with other factors in deciding whether to *Update* or *Redo* the PHA. This is further discussed in Chapter 4.

The quality of the prior PHA is likely to be as dependent on the experience of the PHA team leader in applying the core methodology as it is on the technical and operational knowledge of anyone else on the team. Some companies have programs to ensure PHA team leaders are qualified and remain so, but that is outside the scope of this book. If a company has rules on the qualifications of the PHA team leader, but an exception was granted for the prior PHA leader or the prior PHA leader did not follow the company PHA guidelines (See the discussion of company standards in Chapter 2.), it would justify scrutiny of the quality of the PHA as described in Section 3.2.

3.1.3 *Prior PHA Scope*

The scope of a PHA includes the *physical scope* (i.e., equipment items), *operational scope* (i.e., modes of operation for each equipment item), and *analytical scope* (i.e., consequences of interest). If the prior PHA scope was constrained by any assumptions (e.g., that the hazards associated with raw material and product piping are included in the tank farm PHA), the evaluation should determine whether those same assumptions are valid and applicable to the revalidation.

Physical Scope. It is usually straightforward to verify the physical scope of a PHA by comparing P&IDs or process flow diagrams with the nodes or process sections shown in different colors that correspond to each analysis node or with a separate document that denotes process boundaries. If a company chooses to conduct PHAs on anything less than "all" its process assets, then the physical scope must include all the equipment covered by policy or regulation. In many cases, hazardous raw materials are used to produce non-hazardous products or intermediates. In other cases, non-hazardous raw materials are used to produce hazardous products or intermediates. In the United States, regulators specify covered chemicals and quantities, but a subsequent letter clarified that facilities are expected to establish the physical boundaries of PSM coverage [31]. However, in the referenced letter, OSHA implies an iterative process. Companies should review the results of initial PHA studies to adjust process boundaries. Conservative organizations often extend the physical scope of the PHA at least one (and sometimes two) process sections beyond the covered equipment to ensure the hazards that caused the process to be covered are no longer present, even if all safeguards fail. The physical scope may also include equipment (e.g., transportation containers or vehicles, nitrogen systems, packaged processing units) that is owned by someone else but has an impact on process safety. If the covered process does not begin at a clear physical boundary (e.g., storage tanks, delivery vehicles, or the fence line), the same logic applies to including upstream sections as well.

Operational Scope. Evaluating the operational scope requires more familiarity with the process. Every plant will have startup, production, and shutdown modes of operation. Some plants will have additional temporary or standby modes of operation. Batch plants inherently have different modes of operation during different steps of the production process. For example, a tank may be shown on a P&ID as the "resin prep tank," with little additional explanation. The resin preparation step may simply mix a non-hazardous polymer in water, presenting few or no process hazards. Conversely, the resin preparation step may take days or weeks, and involve flammable or toxic materials to prepare the resin for use, transfer it to the column where it will be used, and prepare the column. In that case, the resin preparation step may require multiple process sections in a PHA. Process hazards and their associated risks may be significantly different during different operating modes. For example, runaway reactions may only be possible during the initial monomer feed step. Thus, it is important to understand what operating modes are possible and which were evaluated in the prior PHA. The typical way to determine whether the prior PHA fully addressed the operational scope is by either (1) interviewing personnel very familiar with the process or (2) carefully reviewing operating procedures. Between these two options, interviews are generally faster and can give more insight than review of documentation alone.

Analytical Scope. The analytical scope of a PHA may include many different types of risks, most commonly categorized as safety risks, environmental risks, and economic risks. Economic risks may be further divided into risks associated with property damage, business interruption, adverse media attention, or poor product quality. Within those broad categories, regulations and company policies may specify consequence thresholds, such as "serious danger to employees," "environmental damage requiring remediation," or "loss of market share."

It is usually easy to verify whether the analytical scope of a PHA meets current regulatory and policy requirements by comparing the stated scope with those requirements. If the "consequences of interest" are not explicitly stated in the prior PHA, they can usually be inferred from the consequences listed in the core analysis worksheets or from the consequence categories of any risk matrix that was used. However, if the consequences of interest are not clearly stated, the reviewer should be alert for other indications that the prior PHA was poorly documented. As discussed in Section 3.1.4, clear documentation of the hazards, risk controls, and consequences of their failure is essential for the *Update* approach to be a viable option.

Beyond specifying the types and thresholds of consequences to be considered, the analytical scope must include all the required topics. PHAs are usually required to address human factors and facility siting issues. The analytical scope might also include other topics, such as loss of essential utilities and services, corrosion damage mechanisms, or the hierarchy of risk controls.

Facility Siting. Specific facility siting issues usually arise during a PHA team's core analysis. In addition, facility siting issues may arise during consideration of external events, such as floods, earthquakes, or accidents (e.g., fires, explosions, or toxic chemical releases) at neighboring facilities. However, facility siting issues may be more comprehensively addressed in complementary analyses in a variety of ways. The CCPS has consistently recommended that PHA teams perform a qualitative evaluation of facility siting issues, such as the example checklists provided in Appendix E. The CCPS has also published a monograph on assessment of natural hazards [32]. Some jurisdictions or company policies may require complementary analyses to identify plausible scenarios and perform detailed dispersion, fire, flood, seismic, and/or explosion modeling to identify facility siting issues.

Human Factors. Like facility siting issues, some human performance gaps, and in many cases specific mistakes (with triggers), are identified in the PHA core analysis. These are often listed as specific causes or factors that can contribute to hazardous events. However, the core analysis techniques do not focus on the underlying causes of human error. For example, teams may list specific actions taken by operators and others to prevent or mitigate the effects of process upsets without considering how robust the human response safeguards are. In such cases, as discussed in Section 4.2.5, the potential loss of knowledge due to staff turnover would be an important consideration. A complementary analysis of general human factors issues, using a checklist such as the examples shown in Appendix F, helps promote more systematic review of factors that contribute to human performance gaps.

Loss of Utilities. The core PHA techniques typically consider the loss of utilities to specific pieces of equipment. A pump may trip due to a loss of electrical power, or a reactor may overheat due to a loss of cooling water. The prior PHA should also consider general losses of utilities and services (electrical power, cooling water, instrument air, nitrogen, steam, fire water, ventilation, and so forth) supporting the process. Some utilities may involve a covered chemical (e.g., chilled water system using ammonia or propane serving as a refrigerant) and warrant a separate node(s) within the PHA or be analyzed in a separate PHA. Regardless, functional failure of these systems should be evaluated as part of a PHA for the unit(s) they support. In addition, loss of support systems not traditionally considered utilities (e.g., the basic process control system, instrumented safety systems, nighttime lighting, the local area network, and communication systems) should be included in the scope. The scope should include consideration of both (1) loss or improper operation of these systems that could cause or contribute to hazardous events in the process and (2) upsets in the process that could cause significant hazards (e.g., contamination, overpressurization) in or via interconnected utility systems.

Typical strategies for addressing these additional topics are discussed in Section 1.2.2. Each of them should be evaluated for quality and completeness as discussed in Section 3.2.

3.1.4   *Drawing Essential Criteria Conclusions*

Prior PHA Methodology. No single core methodology is always appropriate for completing a PHA for every process and situation. The core methodology is often supported by complementary analyses and supplemental risk assessments, all of which comprise the overall PHA methodology. When evaluating whether the PHA methodology used for the prior PHA was adequate, consider:

- Would this PHA methodology be used for a new PHA today?
- Was the PHA methodology selected based on good technical reasons or expediency?

- Was the PHA methodology capable of producing the required results (e.g., a comprehensive listing of relevant process hazards, risk controls, and the range of consequences specified in the scope of the PHA)?
- Was the PHA methodology applied properly? (See Section 3.2.1.)
- Did the PHA methodology document the types of information needed to support the stated PHA objectives?

The answers to these and similar questions provide insight on whether *Update* is a reasonable option for all or part of the PHA; otherwise, a *Redo* may be required. Certainly, if a prior PHA team used an inappropriate core methodology, did not apply it correctly, or did not document essential information throughout, the only option is to *Redo* the PHA. A Checklist Analysis cannot be converted into a HAZOP Analysis by an *Update*. However, if an inappropriate method was only applied to a few nodes (process sections) or operations, it would be possible to *Redo* those nodes and update the rest of the study. If one of the complementary analyses or supplemental risk assessments was deemed inappropriate, the core analysis could be *Updated*, and the inappropriate analysis could be *Redone*. For example, a simple facility siting checklist might be upgraded to a quantitative fire and explosion analysis without *Redoing* the core study.

Prior PHA Inputs. If the prior PHA team had missing or incorrect information that affected the entire process, the *Redo* approach is the best option. For example, a new unit may have been "fast tracked" through the design and construction process, and the initial PHA may have been based on P&IDs issued for construction rather than the "as-built" drawings. If the MOC documentation of field changes during construction is insufficient to determine if process safety risks were effectively managed, the *Redo* approach should be strongly considered.

More commonly, deficiencies in P&IDs or other key documents used by the prior PHA team were the result of factors that have been corrected. The PHA leader will then need to decide if *Redo* is necessary, or if an *Update* is adequate. For example, if a new relief valve and flare study had been done; the revalidation approach might be to *Update* any PHA scenario where a relief device was credited as a safeguard. A combination of the *Redo* and *Update* approaches would be an option if only some of the documents were seriously deficient. For example, if only a few drawings were out of date, analysis of equipment on those drawings might be *Redone* while the rest of the analyses were *Updated*.

If there are significant questions about the experience or knowledge of prior PHA team members, or the PHA leader, the *Redo* approach is likely the best option. If the deficiency is limited to one identifiable issue (e.g., the process control engineer left the company shortly before the prior PHA and no replacement was available during the sessions), it is possible to use an *Update* approach where the team pays particular attention to issues related to that function (in this case, actions taken by the basic process control system and various instrumented safety systems).

Some of the situations described in this section drive the PHA revalidation team leader to go directly to a complete *Redo* approach. If that is the case, the PHA revalidation team leader can skip to Chapter 6 and start preparations accordingly. More commonly, a review of the prior PHA indicates that while methods and inputs could have been better, they were not fatally flawed and can be *Updated*. In those cases, the decision on the best approach is driven by the quality of the prior PHA, extent of changes, and operating experience.

Prior PHA Scope. While failure to address PHA scope items in the prior PHA may preclude using a focused *Update* approach, it does not automatically trigger a full *Redo* approach and therefore, a combination of approaches is usually warranted to ensure the full scope of the PHA is included. It is typically straightforward for a PHA revalidation team to extend the scope of a PHA to address any of these specific items during the PHA revalidation. For example, if it is discovered the prior PHA did not sufficiently address global utilities, the PHA can be *Updated* by adding considerations to the end of each existing node, adding a new node for loss of utilities, or adding a new loss of utilities checklist.

3.2   PRIOR PHA QUALITY AND COMPLETENESS

A revalidation team cannot efficiently fix a fundamentally flawed PHA via any approach other than to *Redo* the core analysis and any complementary or supplemental analyses based on it. However, if the fundamental flaws are in a complementary or supplemental analysis, it (they) may be redone without also *Redoing* the core analysis. Appendix B provides a comprehensive checklist titled "PHA Quality and Completeness Checklist" that can be used for evaluating the prior PHA.

Beyond what was addressed in Chapter 2 and earlier in this chapter, the following sections discuss several additional PHA quality issues. The *Update* approach may not be suitable if the prior PHA is severely flawed due to factors that include:

- Improper application of the analysis method(s)
- Insufficient detail and accuracy of the analysis

- Logic errors and inconsistencies in the analysis
- Failure to document hazards
- Improper application of the risk tolerance criteria

3.2.1   Application of Analysis Method(s)

Chapter 5 of the CCPS book *Guidelines for Hazard Evaluation Procedures* [2] specifies how to apply various PHA methodologies and the results for each method. Chapter 6 in that book covers factors for deciding which method to use in a particular situation. The discussion of PHA methods in this section is limited to the most commonly used core PHA methods: What-If/Checklist, HAZOP, and FMEA. It focuses on their use in the PHA revalidation step (which is the most common form of PHA activity throughout the "operate and maintain" portion of the facility life cycle). It excludes their use for various hazard analysis activities conducted early in the life cycle before the initial PHA (e.g., during preliminary hazard identification or design review activities) or late in the life cycle (e.g., during mothballing, decommissioning, or demolition activities).

Example – Improper Core Analysis Method

A PHA due for revalidation is being reviewed. The reviewer observes that the core analysis is primarily an industry checklist specific to truck/railcar unloading of a hazardous chemical feedstock.

The core methodology used in this study (checklist alone) was likely inadequate. While the checklist performed is useful, it is supplemental to the insights that would be gained using the What-If/Checklist combination or HAZOP techniques.

In practice, management must ultimately decide which analysis method provides sufficiently rigorous information to support their risk judgments and resource allocation decisions. Process complexity and process hazards are often cited as the two main reasons for using more structured methods.

Many view the HAZOP methodology as very deliberate and systematic in structure but more time consuming when compared to the What-If/Checklist method. However, experience shows that the time and effort required to complete PHAs of comparable quality is driven more by the nature of the hazards to be studied than by the choice of PHA technique. That is because PHAs of comparable quality should include the same loss scenarios, identify the same areas of elevated risk, and produce comparable recommendations for risk reduction. In one study [33] published by the American Institute of Chemical Engineers (AIChE), involving six different processes, a total of 414 meeting hours were required when the PHA revalidations were conducted using the What-If method, whereas 198 hours were required for the same six processes in the PHA revalidations that occurred five years later using the HAZOP method. While it is likely that several factors contributed to this productivity improvement, one important factor was changing to the structured approach of the HAZOP method. Focusing the teams' attention on a discrete and organized catalog of deviations was more efficient than having them develop and answer a randomly ordered, long, and often redundant list of what-if questions.

It is important to note that the level of detail applied to any PHA method (and discussed further in Section 3.2.2) can dictate whether its application is appropriate. A skilled facilitator with careful planning and a competent team can complete a What-If/Checklist Analysis that is more appropriate to the process than a poorly conducted HAZOP Analysis, and vice versa.

> **Changing the Core Methodology**
>
> Changing the core methodology (e.g., from What-If/Checklist to HAZOP) for all (or a substantial part of) the process almost always invokes the *Redo* approach.

Assuming management is satisfied with the method and detail used by the prior PHA team, the quality and completeness of the prior PHA should be judged against the expected results from that method.

3.2.2   Level of Detail and Accuracy of the Core Analysis

What-If/Checklist. It can be more difficult to review the completeness of a What-If/Checklist Analysis than other methods, depending upon the degree of structure used in applying the What-If method. A reviewer can usually postulate a question the team did not ask, and those questions can be addressed in an *Update*. However, the *Redo* approach becomes attractive as the review discovers more deficiencies, such as:

- Key questions that were not asked (e.g., no questions about cross-contamination of heat exchanger fluids; no questions about freezing in a cold climate)
- Questions about failures of safeguards instead of the underlying loss scenario (e.g., questions about failure of the sprinkler system rather than questions about potential fire scenarios)
- Responses that implicitly took credit for safeguards in the stated consequences (See example in the following text box.)

If many such deficiencies are found in the core analysis, it is possible that analysis with a different methodology (e.g., HAZOP or FMEA) would provide a more systematic and comprehensive review.

Example – What-If with Insufficient Detail and Accuracy

What-if there was high level in the tank?

If the documented consequence is *"process shutdown,"* the answer is likely taking credit for a high-level shutdown interlock. If the answer is *"overflow to bund,"* the answer is taking credit for secondary containment.

Either way, the responses to the What-If question do not provide a detailed list of safeguards, nor do they accurately describe the consequences assuming failure of the engineering and administrative controls.

When using the What-If method, teams often use larger nodes or process sections, and in so doing, they may completely overlook various process equipment or process operations (unless specifically using a structured variation of this methodology). Therefore, it is important to review both the questions/answers and structure of the What-If Analysis. If the questions asked were too general, or the nodes were too large, hazards may have been overlooked. Consider applying the *Redo* approach as necessary, breaking the existing process into multiple, easier to analyze, nodes and asking questions that are more specific, perhaps on a line-by-line or vessel-by-vessel basis, similar to the HAZOP approach.

When checklists are used for supplemental analyses, they may be organized so the team can summarize the responses for each section of the checklist. This strategy is often used in cases where checklist questions are grouped under headings such as the Human Factors Checklists shown in Appendix F. Summarized/grouped checklist responses do not always mean a Checklist Analysis is lacking in detail, but it may be a reason to look further into the quality of the analysis. The advantage of summarizing responses is it allows the PHA leader to use the checklist as a brainstorming tool.

Example – Summarized Checklist Responses

PHA Team A grouped the human factors checklist questions (See Appendix F.) into topics and had a thorough discussion regarding process controls. Although each question was not specifically answered, the team documented process safety concerns (with recommendations) regarding interlock systems that are normally bypassed and lack of an emergency shutdown panel.

PHA Team B answered each human factors checklist question with the same response: "No issues."

In these examples, even though Team A did not document the answer to every individual question, its analysis appears more accurate and complete than Team B. It is entirely possible that Team B completed a thorough Checklist Analysis. However, Team B's documentation would be more persuasive if it included some description of the engineered or administrative safeguard(s) that resulted in their "No issues" determination. (See also the discussions of PHA documentation in Chapter 8.)

Evaluating the answers to lists of questions is relatively straightforward. When reviewing the documentation from What-If/Checklist analyses, consider:

- *Were all the questions answered as intended (i.e., are there question-by-question or summary responses)?* Failure to respond to a question is an omission.

- *Are the answers detailed enough to provide insight and justification for the response?* Sometimes a simple yes, no, or not applicable response is appropriate, but that should not be the norm.

- *Do the responses make sense from a technical perspective?*

- *Has the RAGAGEP underlying the checklist changed significantly since the checklist was last used?* Outdated checklist questions could solicit outdated responses.

- *Are there cases where elevated risks were identified, but the team did not include a recommendation?* Perhaps the team judged the elevated risks to be ALARP, but the reasons for that determination should be evident.

Note: The suggestions for evaluating HAZOP deviations discussed in the next topic are also germane. If documentation issues are identified, and particularly if they are common, consideration should be given to *Redoing* the analysis. In fact, it is common for teams to *Redo* checklists even when using the *Update* approach elsewhere. For example, if an *Update* primarily focuses on incorporating MOCs into the PHA, the fact that field labelling has degraded due to weathering may be overlooked. One way to ensure the revalidation team considers the current condition of labelling is to *Redo* the human factors checklist.

HAZOP. The HAZOP method uses deviations created by combining guidewords (high, low, no, other than, etc.) with process parameters (flow, level, temperature, pressure, composition, etc.). Questions that should be asked when reviewing the quality of HAZOP studies include:

- ***Are the nodes or sections defined in a manner that deviations have clear meaning?*** If the HAZOP worksheet is quite complex, it may be that the prior PHA team used process sections that were too large. Therefore, each deviation had to be repeated multiple times to ensure all relevant hazards were addressed (e.g., High pressure in Vessel A, High pressure in Vessel B, High pressure in Vessel C). If the reviewer sees such repetition in a very large node, they should carefully look for instances that were missed. Conversely, deviations may have been misunderstood because the process sections were too small. In such cases, the causes, consequences, and safeguards may be divided among so many small nodes that the important equipment and system interactions were overlooked.

- ***Have valid deviations that do not result in a consequence of interest been deleted, or were those rows left blank?*** Documenting which deviations do not cause issues of concern, and why they are not of concern, is also valuable information for subsequent PHA revalidation teams to consider. There should be an entry in each row. If not, the PHA revalidation team will need to address this issue by *Updating* those deviations.

The deviations within the HAZOP worksheet can be sampled to evaluate the level of detail and logic of the analysis. Note: The questions listed apply to the What-If/Checklist and FMEA methods as well, with minor differences in column titles for each method. Relevant questions include:

- *Are causes meaningful?* A cause such as "operator error" for a "High level" deviation is not meaningful to anyone outside of the PHA team, and its specific meaning will probably be forgotten by the PHA team members themselves a few days after the PHA meetings are complete. A more meaningful statement of the cause might read, "operator fails to remain with the tank during the fill step and stop the flow at the 'full' mark," or "operator fails to stop the flow at the 'full' mark due to being distracted by another process alarm." It is easier to understand and revalidate causes that are more specific.

- *Do the stated consequences presume failure of all safeguards?* Taking credit for safeguards in the consequence definition is a common PHA flaw. If this error is persistent throughout the PHA, it is a sign the PHA team did not accurately document and discuss worst credible consequences in the PHA. Consider the possibility of high pressure in a vessel. If "process shutdown" was listed as the consequence, the prior PHA team may have assumed the high-pressure interlock would shut the process down as designed and disregarded the potential for catastrophic vessel overpressure. If "discharge to the flare" was listed as the consequences, the prior PHA team likely assumed the pressure relief valve worked as intended to protect the vessel. PHA teams may be willing to accept the risk of a particular outcome based on some (but not all) of the safeguards working, but if a qualitatively worse outcome went unrecognized, risk may not be managed as intended. In the case of vessel overpressure, the unacknowledged consequences are that the vessel might rupture (if the maximum allowable working pressure [MAWP] exceedance is great enough), and the prior PHA team did not document its implicit acceptance of that risk.

- *If a consequence of interest might result from successful operation of a safeguard, is it addressed as well?* The March 2005 explosion at a Texas City refinery [34] did not involve failure of all safeguards. As the pressure in a distillation column rose, the pressure safety valves functioned as intended, discharging flammable liquid above its normal boiling point to the atmospheric blowdown stack and releasing a large flammable vapor cloud. Ignition of the cloud led to an explosion that killed 15 and injured 180 at the refinery. Similarly, the fire protection systems worked as designed to suppress a 1986 agrichemical warehouse fire in Switzerland. The firewater runoff carried highly hazardous chemicals into the nearby Rhine River and poisoned aquatic life downstream to the North Sea [35]. Thus, it is important that the prior PHA team also considered scenarios, particularly those involving a loss of containment, that result in consequences of interest even if some of the safeguards work properly. Those scenarios should be listed and ranked separately from the scenario assuming failure of all the safeguards.

- ***Is the set of safeguards reasonably specific, relevant, and complete?*** If the prior PHA team listed only vague safeguards such as procedures, training, or maintenance, then it is virtually impossible to revalidate the accuracy of their risk judgments. If the issue is widespread, it is usually more efficient for the team to *Redo* the PHA than guess at the previous logic and edit all the entries as needed to add necessary details. Conversely, a more specific safeguard, such as "High level interlock" can be revised, if needed, to be more specific, such as "High level interlock LSH-406 with action to close feed valve XV-101." Sometimes the listed safeguards do not appear to be relevant to the particular deviation, so at least the *Update* approach will be required to ensure safeguards are appropriate and documented as such. Finally, the listed safeguards should be compared against those shown on the P&IDs. If there were no recommendations to restore the functionality of the unlisted safeguards, then it is likely that the prior team overlooked them or made a simple typographic error. However, if the prior team judged the risk tolerable without them, those oversights can be corrected with an *Update* approach.

- ***Do the recommendations seem reasonable?*** Are there cases where the risk appears to be quite high or RAGAGEP does not appear to have been followed, and there is no corresponding recommendation?

- ***Does every deviation have recommendations?*** When most or all deviations have recommendations, it is a very strong signal that the HAZOP was documented "by exception." This is a common style of documentation for design hazard reviews. The reviewers do not expend resources documenting where the designers have achieved tolerable risk. They are trying to direct the designers' attention to those areas that need further risk reduction. Hence, their sole focus is on recommendations, and the documentation supports their rationale for the recommendations. This style of documentation may be appropriate for an engineering design review, but it never meets regulatory requirements (and rarely meets corporate requirements) for a PHA. If the prior PHA was only documented "by exception," the *Redo* approach is almost certainly required.

FMEA. All the issues identified in the preceding discussion of the HAZOP method in Section 3.2.2 also apply to the FMEA method (ignoring minor terminology differences, such as "effects" instead of "consequences"). The unique concern about PHAs using the FMEA method is whether they adequately considered scenarios involving issues other than discrete component failures, such as, human errors, variations in raw material chemical composition/properties, or other non-equipment related issues. If the FMEA only lists specific component failures as causes, such as "valve fails open," the *Redo* approach is likely required, perhaps using the What-If/Checklist or HAZOP method.

3.2.3   *Logic Errors and Inconsistencies in the Analysis*

Information in PHA reports should be valid, accurate, and logical. To some degree, the validity can be evaluated by having persons familiar with the process review the report. Accuracy can be checked (to the degree details are documented in the prior PHA) by comparing equipment tag numbers and other parameters (e.g., set pressure on a relief valve) to the process knowledge documentation. Logic errors are best discovered by reading the hazard analysis worksheets. The decision to *Redo* or *Update* the PHA depends on the prevalence and magnitude of these errors in the prior team's risk judgments. Examples of logic errors include:

- *Listing human error as a cause with no further explanation.* Trevor Kletz is often credited for the idea that "blaming accidents on human error is equivalent to blaming falls on gravity" [36, p. 2]. Humans will make mistakes. PHA teams should explore reasons that an error might occur (i.e., an error-likely situation) and result in substantial consequences. If feasible, the PHA team should have made recommendations to substantially reduce or eliminate the risk of consequential, unrecoverable human errors.

- **_Listing written procedures and training as the sole safeguard for a human error._** While written procedures and training are very important, taking safeguard credit for them assumes that humans will think, diagnose, and react the same way each time in similar situations. However, these safeguards have already failed if the error occurs. Unless an alarm or some other means alerts the operator to the error, written procedures and training are of little use because the operator is unaware of their mistake and would typically be moving on to the next task.

> **_Example – Procedures and training as Safeguards_**
>
> An experienced operator fills a tank flawlessly on Monday but makes an egregious error and overfills the tank on Tuesday. The procedure was the same, and the operator did not become less trained overnight. Perhaps the operator did not sleep well on Monday night.

- **_Claiming the same item as the sole cause and the sole safeguard._** Claimed safeguards provide no significant risk reduction unless they are largely independent of the causes. The concept of independence is discussed in depth in the CCPS book *Guidelines for Safe Automation of Chemical Processes* [37, pp. 16-17, 372-375]. However, when reviewing the prior PHA, it is sufficient to recognize that alleged safeguard(s) that could be incapacitated by the same failure or error that caused the process upset would typically fail the independence criterion.

> **_Example – Cause as a Safeguard_**
>
> If the prior PHA lists "Malfunction of TE-123" as the sole cause of high temperature and lists "TE-123 with high temperature alarm" as the only safeguard, there is no real/effective safeguard.

- **_Identifying hazards inconsistently._** For example, in a HAZOP, the deviation "No pressure" or "Reverse pressure" might be used to explore the consequences of vacuum conditions. However, if the prior PHA team used those deviations interchangeably, or if they were not used consistently (e.g., No pressure means no gauge pressure; Reverse pressure means vacuum), then it will be very difficult for the revalidation team to follow their predecessors' logic. At the very least, the prior PHA will need to be *Updated* for consistency.

- **_Claiming multiple safeguards that are not independent._** Claimed safeguards provide no significant risk reduction unless they are largely independent of each other [37]. However, when reviewing the prior PHA, it is sufficient to recognize alleged safeguard(s) that could be incapacitated by the same failure or error would typically fail the independence criterion. The choice of _Redoing_ or _Updating_ the PHA depends on the prevalence of this error in the prior team's risk judgments.

> **_Example – Safeguard Independence_**
>
> If the prior PHA lists a safeguard "TE-123 with high temperature alarm" and a second safeguard "TE-123 with high temperature shutdown," these two safeguards are not independent. If TE-123 fails, both safeguards will fail.

- **_True statements leading to illogical conclusions._** Consider the following true statements: (1) High flow in the outlet line from a supply tank could cause a low level in the supply tank. (2) Low level in the supply tank could cause low flow in its outlet line. Possible illogical conclusions: (1) High flow in the outlet line causes low flow in the outlet line, (2) Opening the level control valve too much causes low flow in the outlet line. If the prior PHA contains circular reasoning of that sort, the _Redo_ approach may be necessary.

3.2.4   Failure to Document Hazards

PHAs are generally required to identify the hazards of the process, and those are usually evident in the consequence descriptions. In some cases, the consequences of a deviation, question, or failure mode are simply listed as "release of process material." The prior PHA team may have known exactly what that meant. Subsequent PHA revalidation teams will likely struggle to leap from this generic statement to how people may be harmed (or what other undesirable things might happen). Process hazards are properties of the materials and

> **_Causes are Not Hazards_**
>
> It is impossible for a PHA to list every conceivable cause related to a particular hazard. For example, a PHA might identify "pump stopping" as the cause for a low flow hazard and list "spare pump" as the safeguard. The prior PHA did not need to identify and evaluate every possible reason for the pump stopping (e.g., broken shaft, shorted motor, seized bearing) if the listed safeguard (spare pump) is equally effective for all those cases.

energies present, and their uncontrolled release results in consequences of interest such as a toxic exposure, vapor cloud explosion, pool fire, and so forth. (Any release that could credibly result in a vapor cloud explosion could also result in a flash fire, and it is sufficient to document the worst credible consequence for each class of hazards.) If a risk matrix was used as the basis for risk judgments in the prior PHA, then consequences should be stated in the same terms used on the consequence axis of the risk matrix. Otherwise, consequences should be stated in standard, consistent terminology (e.g., "a flammable release with the potential to affect anyone within battery limits," or "potential off-site environmental impacts requiring remediation"). If the reviewer cannot easily confirm the severity ranking used in the prior PHA by reading the consequence description, then *Updating* the analysis worksheets will be much more difficult.

3.2.5   *Improper Application of Risk Tolerance*

Many companies require PHA teams to risk rank each What-If Analysis scenario, HAZOP deviation, or failure mode that leads to a consequence of interest, and they communicate the organization's risk tolerance via a risk matrix. At a minimum in such situations, PHA teams are tasked with categorizing the severity of unmitigated consequences and categorizing the scenario's mitigated likelihood (i.e., the cause frequency adjusted downward to take credit for any safeguards) on the frequency axis. The intersection of unmitigated severity and mitigated frequency on a risk matrix provides an overall risk ranking that determines whether further actions to reduce risk are needed (i.e., PHA team recommendations if the risk is not ALARP).

To decide on a revalidation approach, the prior team's risk judgments should be examined. They may have been inconsistent – the likelihood of a drain valve being left open on one pipeline might be categorized differently than a similar valve being left open on another line, for no apparent reason. They may have documented the risk rankings "by exception" – perhaps the team only risk-ranked some scenarios and ignored others, or perhaps the team only documented their risk

> **Risk Judgment**
>
> Although it is possible for a fire to lead to a first-aid event (possibly a minor blister?), it is also credible that someone exposed to burning process material could suffer a serious or fatal injury. For more information on establishing and applying risk tolerance criteria, see the CCPS book *Guidelines for Developing Quantitative Safety Risk Criteria* [42].

ranking of scenarios for which there were risk reduction recommendations. The implication is that the risk of other scenarios is tolerable, but there is no way to understand their reasoning.

In some cases, it may appear that the prior PHA team "reverse-engineered" risk judgments. For example, the PHA worksheets may have documented that a release of flammable material would lead to a first-aid/minor injury (based on the worst credible consequence), and similar severity rankings were consistently applied throughout the core analysis worksheets. Upon questioning, the prior PHA team leader might admit that a higher severity ranking would have led to more work (e.g., a requirement to perform a LOPA), and that would have likely led to even more work (recommendations for additional risk reduction measures). Thus, they deliberately assigned a lower severity and/or frequency category to get a lower risk ranking and avoid requirements for more detailed analyses or recommendations. This is symptomatic of a poor process safety culture and its pervasive negative effect on PHA quality. If a reviewer finds several cases where unmitigated consequences were clearly understated, regardless of the prior team's motives, the *Redo* approach may be warranted with a substantially different team.

In some cases, the methodology and analysis are sound, but the risk ranking is inconsistent. For example, the prior PHA team might have thoroughly documented the worst credible consequence of a fired heater accident as an explosion hazard and listed several safeguards, but only ranked the risk as "Low – Tolerable." However, if the site has a history of explosions in heaters with similar safeguards, the likelihood of severe injury or fatality due to this scenario is much higher. Thus, in this case, the risk ranking is inconsistent with actual facility experience. It is possible to address this issue by reviewing each deviation/scenario to confirm or *Update* the risk ranking.

3.2.6   *Drawing Quality and Completeness Conclusions*

Most of the issues identified in this section point to the need to either (1) *Redo* the PHA or (2) use a combination of *Redo* and *Update* approaches. If issues are identified in multiple categories, it is likely that the team will be discussing every scenario/deviation and in that case, it is often more efficient to use the *Redo* approach. It is possible to discover isolated issues scattered among several of the categories identified in this section that can be fixed using the *Update* approach; however, this is rarely the case.

3.3   PRIOR PHA TOPICS FOR ADDITIONAL EVALUATION

In addition to the factors listed in Sections 3.1 and 3.2, as well as Chapters 2 and 4, a few other items to consider when evaluating the prior PHA are addressed in this section.

3.3.1   *Status of Prior PHA Recommendations*

Recommendations from the prior PHA should be resolved before the revalidation, but occasionally, due to their complexity or difficulty of implementation, some recommendations may still be unresolved. The recommendations that were resolved were either accepted or rejected. Regardless of their status, all recommendations should be considered during preparation for the revalidation meetings:

Accepted Recommendations. An accepted recommendation usually results in some change to the process equipment or procedures. Thus, in most cases, there will be a corresponding MOC evaluating the hazards of the change, but not always. For example, there may be no explicit MOC for resolution of recommendations for "training." The reviewer should look for the implementation of those changes when evaluating the operating experience (as discussed in Chapter 4). If no MOC can be found, or if it appears that the actions taken may not have adequately accomplished the intent of the prior team, those recommendations should be highlighted for reconsideration by the revalidation team.

Rejected or Declined Recommendations. If a prior PHA recommendation was rejected or declined, is there any evidence that the prior PHA team was involved in that decision? Perhaps the prior team worded the recommendation poorly and it was rejected because management simply did not understand its value. Another possibility is that the prior team made a technical error of which the upcoming revalidation team should be aware. Perhaps previous management was willing to accept an elevated risk of that particular hazard and documented this decision. Regardless, rejected recommendations should be given special attention by the current revalidation team to verify their concurrence that no further action is warranted.

Unresolved, Open, or In-Progress Recommendations. Any unresolved, open, or in-progress recommendations from the prior PHA should be specifically reviewed during the revalidation. If the team concurs that the risk is elevated, it should reaffirm the recommendation in the revalidation report. If the team concludes that the risk is tolerable with no further action, then the team should provide management with the information needed to resolve the issue.

3.3.2   Complementary Analyses and Supplemental Risk Assessments

Beyond the core methodology used in the analysis, the reviewer should consider the implications of the revalidation on the required complementary analyses (human factors, facility siting, dust hazard analyses, etc.). For example, the facility siting topic is often also addressed in a separate study of the entire site. What, if any, loss scenarios in the facility siting study were extracted from the prior PHA for this analysis, and how will any revisions during the revalidation be captured for inclusion in it? Similarly, many organizations require supplemental risk assessments, such as LOPA or bow tie analysis, to be performed. If LOPA is applied to one or more scenarios in the core analysis (e.g., HAZOP), then the revalidation team is typically tasked with *Updating* or *Redoing* the LOPA, as appropriate, as they revalidate the PHA. However, some organizations commission a separate LOPA team, and in that case, the revalidation team should identify any loss scenarios in the prior PHA that were evaluated in the LOPA. If those scenarios are changed during the revalidation, or if new scenarios are identified, they should be captured for inclusion in the next LOPA revision.

Although the discussions in Sections 3.1 and 3.2 have been focused on the core analysis, the complementary analyses and supplemental risk assessments can be evaluated in a similar manner: (1) obtain and review the current requirements for the analysis, and (2) evaluate the prior analysis to determine if it is significantly deficient or unacceptable.

3.3.3   Opportunity to Learn and Capture Information

Occasionally the planned date for a PHA revalidation coincides with significant organizational turnover. Perhaps most of the senior operators were hired during plant startup 40 years ago, and some of the technical personnel who joined the company right out of college in the same timeframe, are all expected to retire within the next five years. In addition to the more experienced personnel who would normally participate in the PHA revalidation, the team might be expanded to include less experienced personnel. As they either *Redo* the PHA or at least revisit each scenario/deviation, the team can discuss what might happen in each instance (and *Update* the PHA along the way). Thus, the revalidation activities become an opportunity to train less experienced personnel, communicate important information, and if appropriate, better document it in the operating manual or operating procedures for the process.

3.3.4 Continuous Improvement

Some companies depend on having a fresh set of eyes periodically look at their hazard analysis worksheets to both identify (1) previously undetected hazards or unrecognized loss scenarios and (2) opportunities to improve reliability, availability, throughput, yield, and product quality. Although many companies focus primarily on hazards leading to injuries or major loss events, if desired, the PHA can identify operational improvement and equipment reliability opportunities as well as EHS-related hazards.

3.3.5 PHA Documentation Software Changes

Most PHAs are documented using electronic tools. The software falls generally into one of three categories:

- Commercially available, general-purpose software (e.g., Microsoft® Office products such as Excel® or Word®)
- Commercially available PHA documentation software
- Proprietary or company-specific software designed for documenting PHAs

Converting from one software package to another can be challenging (though some software packages offer import functions). If the software is no longer available, no longer provides the required functionality, or is no longer licensed or used by the company (such as when a company standardizes on one software and drops the licenses for others), the revalidation team faces a choice. It can either (1) recreate the prior analysis in the software now being used and then proceed with an *Update*, or (2) recreate the analysis directly as a *Redo*. The most beneficial approach largely depends on the cost and availability of a person or method to re-enter old data, versus devoting those resources to a *Redo* that might yield additional benefits.

3.3.6 Time Since the Previous Redo

Many companies limit the number of times a PHA can be simply *Updated*, and they require a *Redo* every other or every third revalidation cycle (i.e., every 10 or 15 years), regardless of the quality of the prior PHA. One reason is that they want to help ensure new information and perspective are introduced to the hazard analysis process. Another reason is that some changes may not be captured by the MOC process, particularly subtle or creeping changes, and if the PHA is always *Updated*, any associated hazards may never be recognized. For example, if the maintenance department had decided to stop function testing interlocks that stop pumps in low/no flow situations (e.g., based on horsepower readings), the revalidation team may want to re-examine all low/no flow scenarios. Unless the MOC for the authorization to stop testing is associated with every process unit in the facility, it is very possible this change would never come up in the PHA revalidation for a particular unit (unless a previous incident of continued operation of the pump with no/low level in the tank supplying the pump led to pump seal failure).

Yet another reason is that the overall practice of PHA evolves with better tools, better methods, better data, more consensus on assumptions, etc. Practitioners also improve their individual skills, and given the opportunity, many would recommend *Redoing* PHAs from earlier in their careers. The introduction of supplemental tools, such as LOPA, has improved the consistency of PHA team risk judgments. Some insurance carriers consider the age of the PHA in their underwriting decisions. Repeated PHA *Updates* often result in documentation that is difficult to read and use as the basis for MOCs. Repeated PHA *Updates* also tend to perpetuate any analysis deficiencies or omissions. Thus, though difficult to quantify, the benefits of *Redoing* a PHA tend to increase with each revalidation cycle.

3.4   PRINCIPLES FOR SUCCESSFUL PRIOR PHA EVALUATION

Actual experience in conducting revalidations, by PHA practitioners across a number of companies, has highlighted some keys for success, as well as some things that can impede success. While none of the items listed are absolute rules, they do provide valuable guidance.

Successful Practices:

- Developing a written plan for the revalidation *Update* or *Redo* describing the purpose, scope, objectives, and proposed participants, and having it approved by the responsible manager of the process (or project) prior to convening the study meetings
- Evaluating the quality of the prior process knowledge before the PHA revalidation to allow time for correction of deficiencies (e.g., if a field verification of the P&IDs is needed)
- Verifying current, accurate information about all the equipment used in the process including packaged units, leased equipment, shipping/receiving containers, and shipping/receiving vehicles

- Conducting interviews with prior PHA team members to discuss the quality of the prior analysis, along with developing a short checklist of questions for prior PHA team members to help the reviewer gather the information consistently and efficiently
- Conducting interviews with various operations and staff personnel to assess if they believe the prior PHA was adequate
- Adding detail to the PHA responses to have both a more complete record of the analysis and to reduce the time and complexity of future revalidations
- Keeping an edited copy of the PHA worksheets up to date, based on changes made to the process since the most recent PHA
- Having a subset of the PHA team review all safeguards to verify each safeguard is still in place; functional; and periodically calibrated, tested, or exercised
- Involving process experts (e.g., Process Controls, Mechanical Integrity) in the evaluation of the quality of the prior PHA
- Watching for significant gaps in prior PHA documentation (such as missing equipment or PHA nodes)
- Paying attention to the PHA scope boundary and ensuring all the process is covered in the scope of this PHA or another PHA
- Having experienced personnel who were not involved in the PHA meetings perform quality assurance (QA) reviews of PHA documentation
- Addressing the PSM issues that caused deficiencies in the prior PHA outside of the revalidation process

Obstacles to Success:

- Trying to *Update* a PHA without performing an evaluation of the prior PHA first
- Relying on a prior PHA that was completed in a very few meetings. A PHA with substantial scope that is covered in one or two sessions could be lacking in detail or completeness
- Using a prior PHA that was performed only to get it done as quickly as possible and "check the box" of a requirement, with minimal or no effort to perform detailed analysis or develop recommendations for improvement
- Relying on a prior PHA where the team and/or facilitator had inadequate expertise

- Relying on a prior PHA with inadequate information (e.g., incomplete or inaccurate P&IDs or operating procedures)

- Relying on a prior PHA with no node or process section descriptions

- Relying on a prior PHA where the team failed to identify or document all credible hazards associated with the process, capture important initiating events, or address all operating modes (e.g., startup, shutdown, clean outs, catalyst changes)

- Using a prior PHA where there were fundamental issues with application of the core methodology. For example, not carrying consequences to their final conclusion (e.g., stopping at "high pressure in vessel" rather than considering the possibility of "vessel rupture" and its subsequent consequences) or not following loss scenarios to their conclusion in equipment or processes beyond the physical boundaries of the PHA being evaluated

- Using a prior PHA where the team risk judgments were suspect. For example, considering ineffective or unclear safeguards when evaluating likelihood (e.g., taking credit for operator intervention, when in reality the event would develop too rapidly for the operator to respond), or failing to specifically identify engineering or administrative controls (e.g., listing "high level switch" rather than an instrument tag number or "operating procedures" rather than a procedure number)

- Using a prior PHA that had been documented by exception (i.e., only those deviations or scenarios for which severe consequences were deemed likely or resulted in a consequence of interest were documented, such that the reviewer cannot tell whether other scenarios were overlooked, or if they were considered and discounted)

- *Updating* a prior PHA where the previous intention was to not have any recommendations or a prior PHA with an improbably small number of recommendations indicating, perhaps, a too-cursory analysis

- Using a prior PHA with no or limited documentation (e.g., there is no explanation of the methodology used by the prior PHA team)

- Failing to consider future PHA revalidation needs when documenting and implementing MOCs between revalidation cycles, resulting in increased effort for the revalidation team and potential for unexpected results

4 EVALUATING OPERATING EXPERIENCE SINCE THE PRIOR PHA

Two goals of a PHA revalidation are to (1) ensure the PHA is consistent with the current state of the process and (2) ensure the risk judgments are accurate for the current state of the process. Thus, it is essential that the operating history be evaluated for factual input to both the revalidation and the selection of an appropriate revalidation approach (i.e., *Redo*, *Update*, or both). As illustrated in Figure 4-1, this chapter focuses on ways operating history could influence the revalidation analysis approach; Chapter 6 discusses how the collected data from this Chapter could be organized and provided to the revalidation team.

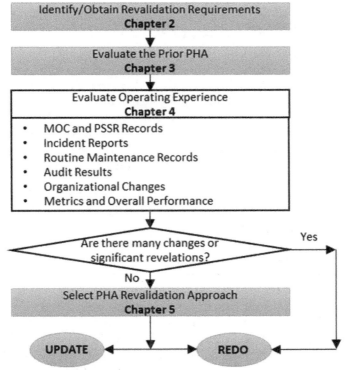

Figure 4-1 Revalidation Flowchart - Evaluating Operating Experience

4.1   OPERATING EXPERIENCE INFLUENCE ON REVALIDATION

How could operating experience affect the approach to the revalidation? Consider two different situations involving a simple ammonia storage tank at a shipping terminal. In Case 1, the tank had been routinely storing ammonia for years, as designed. There had been no changes, no incidents, and no out-of-specification results or unexpected corrosion mechanisms detected during any of its scheduled tests or inspections. In Case 2, the same tank had been modified several times in response to multiple leaks and overfill incidents, and most inspections and preventive maintenance had been deferred due to financial pressure. In Case 2, the risk judgments made by the previous PHA team should be re-evaluated in light of (1) the multiple incidents experienced in the process and (2) changes in maintenance practices that may have allowed the basic integrity of the tank to degrade or rendered critical safeguards ineffective. Therefore, in Case 2, consideration of the operating experience favors the *Redo* approach. Conversely, in Case 1, nothing in the operational experience warrants *Redoing* the PHA, so the *Update* approach should be satisfactory.

4.2   TYPES OF OPERATING EXPERIENCE THAT SHOULD BE CONSIDERED

All operational experience should be considered in preparation for a revalidation, including operational experience with other similar units at the same site, at other sites, and elsewhere in industry, if known. Operational experience broadly includes startup and shutdown activities and any special operating activities (e.g., regeneration of catalyst beds, cleaning between batches). The experiences from unit turnarounds and maintenance activities should be considered as well.

In a few cases, prior operational experience may not be relevant and can be excluded. For example, a batch reactor may have once been multi-purpose, but is now in dedicated service because of increased customer demand for one of those products. Operational experience with other product lines may have minimal or no relevance to the current revalidation.

The following sections discuss several of the most pertinent sources of operational experience. While specifics regarding the selection of PHA revalidation approach (*Redo* or *Update*) are contained in Chapter 5, the sections below are written assuming the default approach to the revalidation is to *Update* the PHA. Wherever possible, the prior PHA will be used as the baseline for the revalidation, and it will be *Updated* to incorporate changes and experience gained since the last PHA. The discussion explains how experience gained since the prior PHA can affirm that the *Update* approach will be sufficient, or if a *Redo* to address certain errors or deficiencies would be more prudent. If significant deficiencies are found in the operational experience records, management should be advised of the issues so they can be addressed outside of the revalidation process.

4.2.1   MOC and PSSR Records

When considering operational history, the number and extent of changes since the prior PHA will have the biggest effect on the revalidation effort. The study leader's first task is to select a "freeze" date for the revalidation. Changes commissioned before the freeze date are included in the revalidation; those occurring afterward will be included in the next revalidation cycle. Typically, the freeze date is about a month before the start of the revalidation meeting(s), so the team will have reasonable time to prepare for the revalidation without constantly revising documentation for the meeting. Sometimes a natural break in the flow of MOCs makes it easy to select a freeze point. For example, the revalidation might include all changes through the October shutdown; any subsequent changes would be included in the next revalidation. The freeze date should be documented in the revalidated PHA, so the next revalidation team will know where to start its review of operating experience.

Terminology: Management of Change, Pre-Startup Safety Review, and Operational Readiness

The MOC program should have a record of all process changes. There is usually a companion record kept by the PSSR program of inspections performed to ensure that any physical changes are ready to be put in service. The RBPS model expands PSSR to Operational Readiness (OR), which ensures that processes are safe to restart, even when no permanent changes were intended. Thus, the needed operational history records may be found in MOC, PSSR, and/or OR documentation.

In general, the more changes and the more complex the changes that have occurred before the freeze date, the more time PHA revalidation team meetings will require and the more likely a *Redo* will be advantageous. However, the complexity of MOCs cannot be considered in isolation – the robustness of the MOC program is equally important. Specific items to consider include:

Whether an effective MOC program has been in place since the prior PHA. The CCPS book, *Guidelines for the Management of Change for Process Safety* [38] describes an effective MOC program. If changes have been made to the process equipment or procedures without using the MOC program, they should be addressed by the revalidation team. If there appear to be only a few isolated changes that cannot be matched to MOC reviews, those defects can be corrected with the *Update* approach. However, if poor or missing documentation makes it difficult or impossible to identify all the changes that have been made, then the *Redo* approach may be necessary to restore confidence in the quality and completeness of the revalidated PHA. Section 5.2.3 discusses indications that can be used to independently gauge the completeness of the operational history of changes. If the MOC program is weak, even simple changes may warrant *Redoing* the affected node(s). However, if the MOC program is robust, the PHA may be *Updated* with confidence that the hazards of changes, even complex changes, were thoroughly evaluated.

The number of changes. Some companies apply their MOC procedure whenever a manufacturer's model or part number changes, even though the replacement meets the design specifications. Even if no new hazard is introduced, these companies find it best to just have one MOC system to ensure ancillary activities (e.g., drawing updates, maintenance and operating procedure updates, and spare part inventory updates) are properly executed. Thus, a change to a new pressure transmitter that is in all respects equivalent to the old one, except that it is more accurate, requires almost no analysis effort by the PHA revalidation team. The *Update* approach is generally satisfactory for simple changes.

However, as the number of changes grows, so does the possibility that potential interactions were overlooked when the changes were evaluated independently. This is particularly true when several changes were implemented in parallel during a major unit outage, and the various project teams were not perfectly coordinated. For example, one project might resolve a PHA recommendation by installing a pump trip interlock to protect against tank overflow. A separate capacity expansion project may have installed several new equipment items upstream of the charge pump and the tank. The no-flow conditions created by shutting down the charge pump could cause more severe process safety hazards than anticipated in the MOC of either project. In this case, the entire affected system should be evaluated, and significant interactions may warrant a *Redo* approach for the affected nodes.

The complexity of the changes. Adding a new solvent recovery system may only count as one change, but it may be a very complex change, requiring significant analysis by the revalidation team. The hazards of multiple projects during a defined outage, such as a turnaround, may be analyzed under one "umbrella" MOC, and the potential for complex interactions may similarly require significant analysis by the revalidation team. However, even simple changes can have far-reaching effects, for example:

- The decision to introduce a recycle stream could result in unexpected corrosion or fouling in the equipment, or it could result in unexpected flow paths during an upset or shutdown.

- A larger pump installed in one unit might be capable of producing shutoff head pressures that exceed the design pressure ratings of equipment in downstream units.

- An interconnecting pipeline might require protection against overpressure scenarios that could have their initiating event at either end of the reversible pipeline.

- A liquid nitrogen carryover from the supplier's facility unit might cause brittle fracture scenarios during purge activities.

Batch formulation changes, utility system changes or process changes requiring several new pieces of equipment or extensive piping modifications might also introduce hazards well beyond the physical boundaries of the changes. Even simple changes may create a complex situation. When several small changes are considered collectively, there may be interactions that have a more significant impact on process risks. Thus, the more complex the process changes, the more likely that the *Redo* approach is warranted.

The degree to which the hazards associated with changes that have been made were analyzed and how they were analyzed. Continuing the previous example, assume the company did add a solvent recovery process and performed the hazard analysis for the MOC in accordance with their core PHA methodology. So rather than simply document the hazards of the change, they documented a comprehensive analysis of the hazards, risk controls, and consequences of their failure. In that case, the revalidation team can directly add the nodes in the project hazard review "mini-PHA" into the existing unit PHA as they closely review and *Update* the prior hazard analysis worksheets. That allows them to identify any oversights while not having to *Redo* that portion of the PHA. When MOC teams use the same core analysis technique as the PHA, meet the same analysis requirements (See Chapter 2.), and document them in the same format as a PHA, the *Update* approach can be used to efficiently accomplish the revalidation goals. Conversely, if the solvent recovery MOC only considered the "hazards of the change," the PHA revalidation team may need to *Redo* the PHA, at least for the new parts of the process.

The time required for review. In a paper published in *Chemical Engineering Progress* in January 2001, the authors provided guidance on estimating the additional PHA revalidation team meeting time required for evaluating different types of changes [39]. Table 4-1 is a revised version of those published guidelines, considering current revalidation practices as described herein and assuming the MOC was conducted and documented using the same core analysis method as the PHA. As the total time estimated to *Update* the PHA nears the estimated time to *Redo* the PHA, the *Redo* approach becomes more attractive because of its potential additional benefits.

Table 4-1 Examples of Estimating Meeting Time per Change

Category	Example Change
Small (15 minutes or less)	• Change the setpoint on a pressure controller • Replace a carbon steel coupling with stainless steel • Change the sampling frequency on a product tank from once per shift to once per day • Install an additional block valve on a drain line
Medium (15 to 60 minutes)	• Replace an existing gas-fired reboiler with a shell/tube heat exchanger • Install a new fixed firewater nozzle on the west side of the process • Replace a rupture disk with a relief valve on a vessel • Install a 2-inch temporary hydrogen line from the hydrogen header to a reactor
Large (1 to 3 hours)	• Install a new storage tank in the tank farm • Tie-in an atmospheric vent line to a thermal oxidizer • Remove a regulator from a unit's nitrogen supply header • Install a new bypass line around the dehydration and scrubbing system from a reactor
Very Large (more than 3 hours)	• Install an additional process train that includes a line with a pump and two columns • Add a new scrubber tower and associated caustic circulation system

From an operational experience perspective, the most important question to answer is "Why was a change made?" Once that is understood, the closely related, follow-up question is, "Was the change actually implemented as planned and approved?" The MOC and PSSR records should provide the answers to those questions.

Usually changes can be categorized as:

- Resolutions of process safety or environmental recommendations (e.g., from PHAs, incidents, or audits)
- Resolutions of process improvement recommendations
- Temporary modifications or impairments

When deciding on a PHA revalidation approach, resolution of a safety or environmental recommendation alone rarely warrants a *Redo*. For example, if a recommendation called for providing additional protection against overheating a reactor, a high-high temperature interlock to shut off the steam supply could be installed as resolution of the recommendation. This type of change can be handled by *Updating* the safeguards in the reactor node.

The resolution of process improvement recommendations may also be amenable to the *Update* approach if they resulted in relatively few, simple changes. However, as previously discussed, numerous and/or complex changes that were analyzed or documented differently than the PHA may require that the *Redo* approach be applied to the new or otherwise affected nodes.

The status of temporary changes. Temporary modifications should be restored to their original configurations within a specified, approved timeframe. From a revalidation perspective, two situations might occur. One possibility is that a temporary modification might coincidentally be active at the time of the revalidation, such as an experimental run of a new product or an engineered clamp on a leaking pipe awaiting replacement during the next shutdown. In most cases, the revalidation team simply notes that change and the date on which it will be reverted to original operating conditions. The temporary MOC addresses the hazards of that change for the authorized period, and the

---

> **Example - Temporary MOCs**
>
> Company A approved a temporary modification of its heat tracing until the earliest known freeze date for that location. If the restoration work could not be completed by that date, the MOC would require revision and formal approval of an extension via the MOC process.
>
> Since the change is not permanent and hazards are addressed in the MOC, the PHA need not be updated if the revalidation happened to occur during the duration of the change.

revalidation applies to the normal configuration. Assuming that the revalidation team does not identify a flaw in the temporary MOC, it will not affect the revalidation effort.

The other possibility is that a temporary change is active beyond its authorized date, effectively becoming a permanent part of the process. In that case, if the temporary change cannot be removed or properly reauthorized, the revalidation should treat the change as if it were permanent. Then, as with other changes, the revalidation team should consider how thoroughly the MOC evaluated the hazards of the change, given its complexity and potential effects on other systems. The simpler the change and the more thorough the MOC, the stronger the case for the *Update* approach.

Another type of change is a temporary impairment, and these are most commonly associated with critical equipment malfunctions (e.g., in control systems, shutdown systems, relief systems, or fire protection systems). It is inevitable that equipment will fail, and to continue operations, management may approve a temporary or emergency MOC that typically alters routine and/or emergency procedures until the device can be repaired. The core question is, "How often are such impairments required," and this question can be answered by looking at the number of emergency/temporary MOCs or at the number of emergency work orders for the unit. If impairments are brief and infrequent, then they have no significant effect on the revalidation strategy. However, if impairments are frequent, lengthy, and/or chronically overdue for restoration and closure, then it is possible that the risk judgments in the PHA are seriously flawed because safeguards were likely credited based on their availability and effectiveness. Interim control measures, such as operator response to an alarm or operator monitoring of a process parameter, typically have significantly higher probabilities of failure on demand than do the automated process monitoring or trip systems associated with the impaired instrumentation. Frequent, lengthy impairments of a few equipment items can be addressed by *Updating* the affected nodes. However, if many nodes are affected, then the risk judgments throughout the PHA may need to be *Updated* or *Redone*, depending on the extent and severity of the issues.

In summary, there is no simple rule for deciding whether to select the *Update* or *Redo* approach based solely on the number and complexity of changes since the prior PHA. The merits of each change situation should be evaluated on a case-by-case basis because specific situations may dominate the choice of revalidation approach. However, it is equally important that the MOCs be considered holistically, both in terms of their number and complexity, and in the context of the robustness of the MOC program that produced those records when deciding on a revalidation approach.

4.2.2 Incident Reports

Incident investigation reports are an important type of operational experience that will strongly influence the revalidation approach. The term incident encompasses events with actual process safety losses and/or the reasonable potential for process safety losses ("near misses"). Some companies broaden the definition further to include any demand on, or activation of, an engineered safety feature (e.g., a relief valve lift or a deluge activation, performance deficiencies in those systems, or exceedances of safe operating limits). The goal of process hazard analysis is to reduce the risk of process safety losses to tolerable levels, so major loss events should be relatively rare or nonexistent. To supplement the reported incidents, the revalidation leader should solicit the recollections of team members about abnormal situations where any safeguard failed, but a loss was averted. Evaluations of emergency response drills and exercises involving on-site and/or community resources may also provide valuable insights. Lessons learned from major incidents involving similar chemicals, equipment, or processes in industry are often available from reputable sources, such as those listed in Table 4-2, as well as from books such as *Incidents That Define Process Safety* [25].

Table 4-2 Example Sources of Incident Information from Industry

CCPS Process Safety Beacon
CCPS Process Safety Incident Database
European Process Safety Centre
EUROPA – eMARS Dashboard – Major Accident Reporting System of the European Commission
United Kingdom Health and Safety Executive
United States Chemical Safety and Hazard Investigation Board

Regardless of the source, incident reports provide a wealth of information that is useful to the revalidation team. Therefore, when looking at a unit's incident history and contemplating revalidation, answers to the following questions are crucial:

- **Did the PHA identify the loss scenario that occurred?** If not, was that an isolated oversight or indicative that the PHA team overlooked other scenarios? The PHA team may have been unaware of a particular damage mechanism, unfamiliar with actual operating practices, or simply rushed. Regardless, any strong indication of multiple or systemic deficiencies in the PHA virtually dictates the *Redo* approach. Furthermore, any scenario-specific analyses (e.g., LOPA, bow tie) related to previous incidents will have to be *Updated* to include any newly identified scenarios with significant consequences.

- ***Did the PHA team identify the loss scenario, but misjudge the likelihood of the cause(s)?*** The team may have underestimated the frequency of equipment failures or human errors that could result in process safety losses. If true for the actual incident, were the causes of other potential scenarios similarly underestimated? If not, an *Update* can remedy the error; if so, all the risk evaluations may need to be *Redone*.

- ***Did the PHA team identify the loss scenario, but misjudge the severity of the consequences?*** The team may have underestimated the quantity of material that could be released, the likelihood of ignition, or the number of people who could be affected. If true for the actual incident, were the consequences of other potential scenarios similarly underestimated? If not, an *Update* can remedy the error; if so, all the consequence estimates and risk judgments likely need to be *Redone*.

- ***Did the PHA team identify the loss scenario, but misjudge the capability of the safeguards, probability of safeguard failures, or independence of multiple safeguards?*** Did the PHA team list (and rely on) vague safeguards such as "procedures," "training," or "maintenance?" If the actual incident involved deficiencies or breakdowns in the engineered and/or administrative controls, then are the claimed safeguards against other scenarios also questionable? If not, an *Update* can remedy the error; if so, the scenario frequency and risk judgments likely need to be *Redone*.

When applying these questions to a review of incidents in a given process, it is typically the number and/or severity of incidents or PHA-related deficiencies that determines the revalidation approach. Learnings from a single incident or failure can usually be incorporated by the *Update* approach. Systemic issues usually will require a *Redo* of the PHA or at least a *Redo* of the portions involved in the identified issues.

Undocumented Incidents

Just like the review of MOCs may discover changes that escaped the MOC system, the review of incident reports and discussions during the revalidation meetings may identify previous incidents that were not formally investigated. Such discoveries indicate deficiencies in the incident investigation program and may warrant a more extensive search for information about previous incidents. Other records (e.g., operating logs, maintenance logs, and process data historian) could be queried to develop a more complete understanding of previous incidents that the revalidation team should consider.

4.2.3   Routine Maintenance Records

Most routine maintenance records have no bearing at all on the revalidation approach. They are, after all, "routine." Yet, they do offer two key insights.

First, is routine maintenance being performed as assumed by the previous PHA team? Routine maintenance is often given a low priority or eliminated, particularly in times of financial stress. The consequences of forgoing routine inspection, testing, or preventive maintenance, or not performing it on time, are real reductions in equipment reliability. Those reductions are often not apparent until the equipment is needed during a process upset or emergency. If routine inspection, test, and preventive maintenance (ITPM) tasks are not being performed as scheduled (e.g., a large number of overdue inspections) the *Redo* approach may be appropriate. It is likely that the risk judgments in the prior PHA were too optimistic with respect to current equipment failure rates (e.g., higher/incorrect initiating event frequencies and safeguard probabilities of failure on demand [PFDs]). If significant maintenance deficiencies are identified, supplemental risk assessments such as LOPA may also have to be *Redone* because the PFDs may change by more than an order of magnitude (e.g., a Safety Integrity Level [SIL]-2 interlock with reduced routine maintenance might only qualify for SIL-1 performance, or might not satisfy the requirements to qualify for any independent protection layer [IPL] credit at all [37]). Note: ITPM is generally presumed to reduce equipment PFDs, but excessive testing of some types of equipment, such as emergency diesel generators, may have the opposite effect. Equipment that is bypassed or removed for maintenance is not available to respond to an emergency, and it may not be properly returned to service after ITPM is complete.

Secondly, routine maintenance records can also provide insight on failure mechanisms overlooked by the previous PHA team. Consider this example of a simple piping circuit. Routine inspection detected pipe thinning, and it was replaced in-kind. A couple of years later, routine inspection detected pipe thinning, and the same piece of pipe was again replaced in-kind. No MOC was required, because in each case, the pipe was replaced in-kind. Yet the pipe is being replaced at a much-higher-than-expected frequency.

> **Example - Customer Complaints**
>
> The customer complaint database included several records of recent, unexpected problems with a new type of shipping container. This alerted the revalidation team to issues that might also arise when the containers were being filled in the process unit. Ultimately, the revalidation team recommended additional ITPM to reduce the risk of loss of containment events.

Perhaps this particular piece of pipe was improperly specified for its actual service conditions. Perhaps the pipe is subject to higher flow velocities or cavitation and eroding at an unexpectedly high rate. In this example, the affected node can be *Updated* during the revalidation to address the actual operating history. However, additional pipe failures were in different pipe segments, suggests a more widespread damage mechanism unanticipated by the previous PHA, or that the current production rates are exceeding the expected flow velocities throughout the unit. Such evidence in the routine maintenance records could justify the *Redo* approach.

4.2.4   Audit Results

A five-year PHA revalidation cycle will typically include one or two site audits that examined the management systems for PHAs and other relevant elements of the site's process safety management system (e.g., Process Safety Management [PSM], Risk Management Plan [RMP], Risk Based Process Safety [RBPS], Seveso). It is a good practice to look at those audit results for any findings against the particular PHA being revalidated, or against the PHA management system in general. If there were particular findings against the current PHA, those were likely addressed to close the audit. But that should be confirmed by the team. If there were more general findings against the PHA management system, then the study leader should evaluate the extent to which they are relevant to this particular PHA. General audit findings like "PHAs are not documenting consequences assuming failure of all the safeguards," or "PHAs are not using the current risk matrix," if relevant, will almost certainly require the *Redo* approach for revalidation.

4.2.5   *Organizational Changes Not Addressed by MOCs*

Ideally, organizational changes are handled as part of the MOC system discussed in Section 4.2.1. However, if the company does not have a formal procedure for Organizational Change Management (OCM) [40], then hazards associated with these changes may not have been fully considered.

For revalidation purposes, organizational changes can typically be categorized as one of the following three types:

1.   Management structure and the chain of command
2.   Staffing levels
3.   Turnover

Management Structure and Chain of Command. Have there been any changes that would significantly affect the risk assessments made in the previous PHA? Some questions to consider that might affect the revalidation strategy include:

- Have there been any changes in key positions related to process safety (e.g., process safety manager/coordinator, EHS manager, maintenance manager, or lead inspector)?
- Have there been any changes in senior management or the leadership team, such as a new plant manager?
- Have there been significant changes in the operational chain of command? For example, have front-line workers been "empowered" to do their jobs without an area supervisor? Is the unit now part of a different operational group?
- Has the site been involved in a merger, acquisition, or divesture?
- If the unit is in an industrial park, have shared service arrangements been changed? Would any of the changes affect the reliability of utility supply services and emergency response services?
- Have there been significant changes in the labor organization or contracts?
- Have operations, maintenance, or engineering services been moved offsite, outsourced to third parties, or brought back in-house?
- Has operational responsibility or technical support been physically moved away from the unit to a central or remote location?

If the answer to any of these questions is "yes," then past operational experience may not be a reliable indicator of future expectations, and operational experience after the change becomes significantly more important. If the change has had (or is expected to have) minimal or no direct effect on the unit, then the *Update* approach is a viable option; however, if the change has had (or will likely have) a major impact in areas such as central control, maintenance backlogs, or emergency response, then the *Redo* approach will probably be the better choice.

Staffing. A second area of inquiry should explore any changes in the number of front-line workers or their responsibilities. Staff changes may adversely affect human factors related to detecting or responding to process upsets. If staff numbers have been significantly changed since the previous PHA, issues such as the following should be considered:

- Are workers subject to mandatory overtime requirements? Are there any limits on the total hours worked or the consecutive hours worked in a defined time period?
- Are required practices for managing fatigue being maintained?
- Are exceptions to the Fatigue Management policy being handled as before the staffing change?
- What tasks are not being done, or are being done differently, due to the staff changes?
- Is there evidence that field checks of equipment and local instruments are being done as required?
- Are there enough operators to respond to alarms in a timely manner, particularly during major upsets?
- In upset situations, must operators wait for assistance before responding?
- Are there enough operators to walk down job sites before issuing work permits?
- Are the backlogs of maintenance or engineering tasks increasing?
- Has the facility transitioned from 24/7 operator presence to unoccupied remote monitoring on weekends or night shifts?
- Are there chronic personnel shortages that affect operations on a particular shift or day?

The CCPS book *Guidelines for Managing Process Safety Risks During Organizational Change* [40] details a procedure to conduct risk assessments to evaluate impact of reduction of personnel during regular and emergency operations. The PHA leader should challenge the PHA team on the validity of the safeguards or protection layer in light of the staffing changes. If the staffing changes only affect a few scenarios or particular safeguards, then the *Update* approach should be adequate.

> **Example – Staffing Reduction**
>
> A company significantly reduced operator staffing after a state-of-the-art distributed control system was installed. The company believed that this change would improve the overall reliability of operations by automating many of the monitoring, control, and emergency response actions. An *Update* of the prior PHA may be inappropriate as many fundamental assumptions made by the prior team should be questioned.

However, if the staffing changes may have significantly increased the likelihood of human performance gaps causing upsets or delaying the response to many loss scenarios, this may indicate the need for the *Redo* revalidation approach. For example, the temporary staff reduction associated with a coronavirus lockdown reportedly contributed to a major styrene release in Visakhapatnam, Andhra Pradesh, India, in 2020.

Turnover. Staffing turnover may also significantly affect the revalidation strategy, as mentioned among the human factors topics in Section 3.1.3. It is very common for a group of workers to be hired when a new unit is put in operation. That group participates in the turnover of the unit from the project team to the operations group, and some of them will likely stay with the unit for many years. Most importantly, they know the unit's original design intent and its history of changes. For PHA revalidation, the key question is whether that same veteran team is responsible for ongoing operations. Perhaps many have retired or are planning to retire. Perhaps some have found job opportunities elsewhere.

If the current or prospective operating staff lacks the deep process knowledge that was needed in the previous PHA, then the *Redo* approach could be a wise choice for the revalidation. In fact, some companies require a *Redo* in high turnover situations for its training value. They believe that the mental exercise of carefully thinking through potential loss scenarios will improve the performance of their less experienced workers, both by avoiding upsets and by more effectively responding to those that do occur.

4.2.6   *Metrics and Overall Performance*

Identifying and using relevant process safety metrics over the life of a process is one of four elements in the RBPS pillar of learning from experience. Metrics are leading and lagging measures of process safety management efficiency or performance. Metrics include predictive indicators, such as the number of improperly performed line-breaking activities during the reporting period, and outcome-oriented indicators, such as the number of incidents during the reporting period.

When preparing for a revalidation, the study leader should first identify what metrics are available, and which of those are relevant. For example, a company might track metrics related to alarm management. Metrics available on previously discussed topics, such as MOC or incident investigation, can be reviewed.

Other metrics that would be useful include:

- *Process Safety Information/Process Knowledge Management.* What is the status of P&IDs and other key plant/unit drawings? Have the relief valve and flare calculations been revised for recent changes? Is there a backlog of drawing changes awaiting drafting? For example, if the site has a three-month drafting backlog and the revalidation meetings will be held in September, the revalidation may have to use the June design as its basis (and note the June freeze date).

- *Process Hazard Analysis/Hazard Identification and Risk Analysis.* Have all the recommendations from the previous PHA been resolved? Were any rejected? Specific scenarios that warranted those recommendations, whether accepted or rejected, must be included if the *Update* approach is selected so the revalidation team can either affirm the risk is tolerable or make recommendations for further risk reduction as discussed in Section 3.3.1.

- *Mechanical Integrity/Asset Integrity and Reliability.* Is there a backlog of overdue tests or inspections? Is repair work being postponed? Any metrics that indicate the asset integrity and reliability systems are not being adequately resourced are warning signs. The previous PHA team's judgments about the likelihood of equipment failures causing upsets and the probability that engineered controls will fail on demand may be systemically wrong, so the *Redo* approach is favored.

- *Training and Performance Assurance.* Is there a backlog of overdue training? Are job positions not being filled? Any metrics that indicate the training and performance assurance systems are not being adequately resourced are warning signs. The previous PHA team's judgments about the likelihood of human errors causing upsets and the probability that operators will fail to properly respond to indications and alarms may be overly optimistic, so the risk assessments need to be reviewed and *Updated*, as appropriate. If the affected number of nodes or scenarios is significant then a *Redo* may be justified.

- *Conduct of Operations.* Are routine activities being reliably executed? Is good housekeeping being performed? Are signs, labels, and lighting being properly maintained? Any metrics that indicate a breakdown in operational discipline are warning signs. The previous PHA team's judgments about the likelihood of upsets may be overly optimistic, so the risk assessments need to be reviewed and *Updated*, as appropriate.

- *Emergency Management.* Is there a backlog of overdue tests or inspections of equipment relied upon to execute the emergency response plan or procedures? Are emergency drills or exercises overdue? Are tabletop exercises being postponed? Any metrics that indicate the emergency systems are not being adequately resourced are warning signs. The previous PHA team's judgments about the mitigation of losses of primary containment may be overly optimistic, so the risk assessments need to be reviewed and *Updated*, as appropriate.

Metrics can provide leading indicators of performance that are critical to a facility's ability to determine if process safety incidents are likely to occur. The revalidation study leader should make a judgment as to whether any current performance gaps are being corrected or whether they will likely persist. The more widespread the performance gaps and the longer they are likely to persist, the more likely the *Redo* approach is needed to judge current risks accurately.

4.3   HOW OPERATING EXPERIENCE AFFECTS THE REVALIDATION

A review of operating experience since the prior PHA revalidation should be completed as part of the assessment to determine whether a *Redo* or an *Update* will be the more appropriate approach for the upcoming revalidation. Chapter 5 will discuss making this revalidation approach selection in detail, but the operating experience review by itself can necessitate one approach over the other.

These examples illustrate how recent operating experience affects the revalidation approach and preparation.

Example 1 – Significant Incidents:

Background. A refinery crude unit is due for PHA revalidation. A few changes have been implemented since the prior PHA, and the PHA includes LOPA for selected higher consequence scenarios. However, because of a fire associated with the unit heater, a large project has been completed to install a burner management system (BMS), convert from naphtha fuel to natural gas, and convert the heater from forced draft to natural draft.

Review of Operating Experience. As part of this review, the process safety coordinator for the facility decides that most of the unit PHA can be *Updated*. Few changes or incidents have occurred in the unit feed and fractionation sections, and unit mechanical integrity reports show no issues for this portion of the process. However, a decision is made to *Redo* all sections of the PHA associated with the heater, because many complex changes have occurred that were not analyzed by the prior PHA team. In addition, the existing LOPA scenarios associated with the heater should be *Redone* to properly account for the IPLs added or affected by the BMS project. This information is given to the current revalidation leader and team prior to the meeting.

Example 2 – Organizational Changes:

Background. Company A has been purchased by Company B and a significant turnover has occurred in the operations, engineering, and management departments. No personnel who were involved in the prior PHA are available.

Review of Operating Experience. This situation is challenging. A significant amount of operating experience has been lost. Company B should probably *Redo* the PHA for several reasons, including consistency with corporate risk tolerances/procedures and development of process knowledge for newer staff. However, in cases such as this, the prior PHA should not be disregarded and should be used during the *Redo* to help the revalidation team understand the original design and current intentions of the process.

Example 3 – Increased Rigor:

Background. A company's internal audit found that misapplication of LOPA was an issue at several process units within the facility. Conditional modifiers and enabling conditions were overused, and therefore insufficient IPLs were installed.

Review of Operating Experience. As part of this review, a decision is made to *Redo* all the LOPA scenarios for the affected process unit. Since other aspects of the PHA qualify for an *Update*, only the LOPA will be a *Redo*. In this case, the full history is crucial information for the PHA revalidation team and should be communicated prior to initiating the revalidation.

Example 4 – Transient Operations:

Background. A company has experienced multiple losses of off-site electrical power due to wildfires in the region. The local electricity supplier changed its policy and now proactively de-energizes power lines to reduce their risk of causing or contributing to fires. As a result, the company now must manage unit shutdown, restart, and standby operations much more frequently than in years past.

Review of Operating Experience. In this case, the disruptions caused by external factors are crucial information for consideration by the PHA revalidation team. As part of this review, a decision is made to *Redo* all the scenarios involving loss of power to the affected process unit. In addition, the analyses of some transient operations (startup, shutdown, and hot standby) will be *Redone*. All other aspects of the PHA qualify for an *Update*.

4.4   PRINCIPLES FOR SUCCESSFUL OPERATING EXPERIENCE EVALUATION

Actual experience in conducting revalidations, by PHA practitioners across a number of companies, has highlighted some keys for success, as well as some things that can impede success. While none of the items listed are absolute rules, they do provide valuable guidance.

Successful Practices:

- Allowing adequate time prior to the revalidation start date to gather and review operating experience since the prior PHA

- Documenting MOC hazard reviews thoroughly and so they can be easily addressed and *Updated* into the unit PHA. For example:
  - o Using the same software and format to document MOC hazard reviews and PHAs
  - o Ensuring personnel/contractors who perform and approve MOC hazard reviews are familiar with core PHA methods and the PHA requirements of the facility
  - o Prior to determining the revalidation approach, reviewing the MOC change log with experienced operations personnel to ensure that all significant changes have been included in the revalidation scope. In addition, auditing the P&IDs in the field with experienced operations personnel to see if any changes can be found in the field without proper documentation in the MOC system
  - o Keeping a working copy of the prior PHA updated with approved MOCs over the revalidation cycle
- Considering changes in interfacing systems (e.g., upstream units, downstream units, utilities) that could affect loss scenarios in the PHA being revalidated
- Searching incident databases maintained by industry groups and other similar facilities within the organization to help broaden the collection of incidents for the revalidation
- Reviewing metrics relevant to risk judgments during the revalidations, such as alarm system performance, overdue ITPM tasks, temporary impairments, drawing revision backlog, overdue training, and incident trends
- Reviewing any recent PSM (or similar) audits to identify any specific findings related to the PHA being revalidated or general findings related to key elements (PHA, PSI, MOC, mechanical integrity [MI], etc.) relevant to the revalidation
- Addressing the PSM issues that caused deficiencies in the operating experience records outside of the revalidation process

Obstacles to Success:

- Trying to force a revalidation into an *Update* approach when a *Redo* would be the more efficient way to incorporate many complex changes/incidents
- Trying to force a revalidation into an *Update* approach when many of the changes were not implemented as approved by the MOC reviewers

- Performing a *Redo* of a high-quality, complete PHA without considering information documented in the worksheets of the previous PHA report, potentially resulting in the loss of process knowledge and experience embodied in the prior PHA

- Failing to cross-reference MOCs when loss scenarios in one unit (e.g., in utilities) may affect other units. Larger/complex changes, changes performed outside the normal MOC process (i.e., capital projects), or changes that affect multiple process units are not always easy to find in the facility MOC system. Unknown changes may be inadvertently missed in the revalidation or add scope/time to the meetings, especially if there have been potentially relevant changes in parts of interconnected systems that are either upstream or downstream of the process segments within the scope boundaries of the PHA revalidation

- Relying exclusively on MOC documentation to identify changes

- Failing to look beyond the process boundary when there are complex interconnections with adjacent processes or equipment/facilities owned by others. Full understanding of loss scenarios may require discussion between the various stakeholders to properly understand safeguards operating near or at the interface between separate units or transportation vehicles.

5 SELECTING AN APPROPRIATE PHA REVALIDATION APPROACH

As stated in Chapter 1, the primary goal of a revalidation is to verify that the PHA document accurately describes the current risk profile of the subject process unit. While a PHA revalidation can sometimes be accomplished with less effort and in less time than the initial PHA, this is highly dependent upon the scope and quality of the prior PHA and thoughtful planning.

The preparation described in Chapters 2 through 4 of this book, and illustrated in Figure 5-1, were all intended to help the reader answer the following pivotal question:

Which PHA revalidation approach will be the most resource-efficient and effective way to produce a complete, compliant, up-to-date, and thoroughly documented PHA?

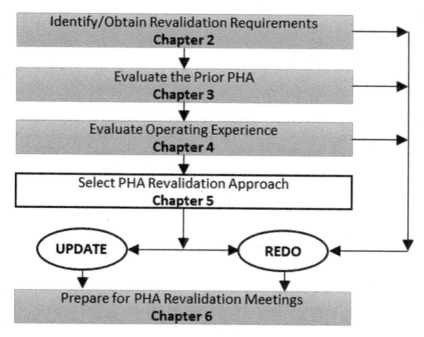

Figure 5-1 Revalidation Flowchart – Selecting the Approach

5.1   REVALIDATION APPROACHES

The two revalidation approaches (introduced in Section 1.5) are described in detail in the following sections.

5.1.1   Update

The *Update* approach is an incremental revision to an existing analysis to reflect the operational experience since the prior PHA was conducted, as discussed in Chapter 4. An *Update* is generally a less laborious option than a *Redo* and is a viable option for high-quality PHAs (as discussed in Chapter 3) of processes at facilities where changes have been effectively managed and few or no significant incidents have occurred. In these situations, the revalidation team can systematically reaffirm the validity of risk judgments documented in the prior PHA and *Update* it with the results from relevant management of change (MOC)/pre-startup safety review (PSSR) reviews and incident investigations.

Two common ways of *Updating* a PHA include:

Detailed Review. The detailed review is similar to *Redoing* the PHA, but should entail less time and effort, because the content of the PHA report is only amended on an "as-needed" basis and not always developed from first principles. Each section or node of the core hazard analysis, as well as any complementary analysis (such as a human factors checklist), in the PHA is reviewed in detail. Changes are discussed to determine their potential impact on previously documented scenarios, or to determine if new scenarios need to be added. (See Section 4.2.1.) Changes implemented in response to prior PHA recommendations or incident investigations are included in this review. (See Section 3.3.1.) They typically result in revisions to the list of safeguards, but they may also affect other portions of the PHA worksheets. Those scenarios affected by the changes, new learnings, or incidents need to be thoroughly addressed, and necessary changes in the PHA are recorded in the PHA documentation (e.g., HAZOP worksheets, Checklist responses.)

Change and Incident Review. Alternatively, the revalidation team may choose to review only the changes and incidents that have occurred since the prior PHA. The revalidation team, with appropriate reference to the prior PHA, discusses the significance of the changes and incidents, and then documents their deliberations and decisions. The revalidation team may use a separate worksheet, and the information recorded usually includes a description of each change (or incident), its process safety significance, and any risk management decisions or recommendations made. (See Sections 3.3.1, 4.2.1, and 4.2.2.) Either

(1) this worksheet alone or (2) the worksheet along with PHA documentation that has been revised per the changes and incidents serves as the *Update* to the PHA. Using only the change and incident review worksheet is sometimes referred to as a "Focused" or "Simple" *Update*.

Regardless of the *Update* approach, the amount of time needed to perform the *Update* will depend upon the extent and number of changes made to the process since the prior PHA, and the number and significance of incidents that have occurred.

When is an Update by "Change and Incident Review" Appropriate?

Facilities should thoroughly evaluate the prior PHA (as discussed in Chapter 3) before initiating an *Update* by Change and Incident Review (Focused *Update*). Because many scenarios documented by the previous team are deemed unchanged and are not being reviewed, (1) the prior PHA must be of high quality and complete and (2) every change and relevant incident must be thoroughly documented and available for the revalidation team to review.

For example, if a process subject to PHA revalidation is an ammonia tank only, with one pertinent MOC, one minor incident, and a high-quality prior PHA, *Updating* the PHA by Change and Incident review can result in a complete, up-to-date, and adequate PHA.

On the other hand, if a process is complex with multiple changes and several incidents, using the Change and Incident Review alone may not produce a PHA that accurately describes the current process risks or may produce a document that will be difficult to revalidate in the future.

Performing a Focused *Update* by Change and Incident Review on an incomplete or otherwise unacceptable PHA can only result in an incomplete or unacceptable PHA revalidation.

Revalidation teams may encounter situations where the prior PHA was of very high quality (i.e., no apparent gaps or deficiencies) and where there have been no significant incidents or changes within the subject process. Such circumstances may permit a much simpler revalidation effort and report, limited to affirming the continued validity of the prior PHA.

5.1.2   Redo

The *Redo* approach is usually selected when an *Update* of the prior PHA is impractical or too complex. This is based on the evaluation of the prior PHA with respect to the criteria as described in Chapter 3 and the evaluation of recent operating experience as described in Chapter 4. In this approach, the revalidation is essentially a new PHA. The team starts from the beginning and performs the entire PHA in detail as if it were an initial PHA. Figuratively, the team starts with blank analysis worksheets, but in practice, most use some information from the previous PHA as a reference or guide.

A *Redo* may be the most appropriate choice in situations where <u>significant:</u>

- Changes have occurred to PHA requirements or analysis methods
- Gaps or deficiencies have been discovered in the process safety information used as the basis for the prior PHA
- Gaps or deficiencies exist in the prior PHA documentation, analysis method, or the manner in which it was conducted
- Changes, particularly uncontrolled changes, have occurred in the process or equipment since the prior PHA was conducted
- Incidents show gaps or errors in the prior PHA

The *Redo* approach involves conducting a PHA of the process using one or more methodologies that are appropriate for the hazards and complexity of the process. The requirements and activities for a *Redo* generally parallel those of an initial PHA. Significant instructional information is available elsewhere (e.g., in the CCPS book *Guidelines for Hazard Evaluation Procedures* [2]) on the conduct of initial PHAs.

Changing the Core PHA Methodology

The decision to *Redo* a PHA offers the opportunity to reconsider the choice of hazard evaluation core methodology to be used (e.g., HAZOP Study, rather than What-If/Checklist).

If a methodology substitution is made, it should be with the intent of enhancing the quality of the PHA, and not at the expense of accuracy or thoroughness. Section 3.1.1 discusses technique selection in light of the nature of the process and its hazards.

Even if the core analysis can be *Updated*, the complementary and/or supplemental analyses may warrant the *Redo* approach. Revalidation team members should be cognizant of any documentation or analysis shortcomings in the prior PHA and strive to ensure the *Redo* remedies those problems.

Example - Overlooked MOCs

Company A has 10 process units at the same site. An MOC was completed four years ago as part of the project to install ambient detectors for toxic gas throughout the facility. That MOC was documented and labeled as a **Unit 1** change because that unit had the highest concentration of the material.

The **Unit 4** revalidation team *Updating* its PHA might easily overlook this change and the unique aspects of the **Unit 4** emergency response that were not considered in the MOC. However, if the revalidation team were *Redoing* the PHA, the discussion should include the appropriate response to toxic gas alarms, even if the MOC was not included in the list of changes.

There may be circumstances where a *Redo* is preferable, not because of gaps or deficiencies in the prior PHA, but from a purely logistical standpoint. For example, if a process has experienced a large number of changes (or very significant changes) since the prior PHA, *Redoing* the PHA may be more time- or resource-effective than *Updating* the documentation to incorporate each MOC. In other words, there may be so many changes to a PHA that starting over from the beginning will take less time and result in a higher quality PHA than searching for each change and ensuring it is adequately documented. A similar situation may arise if the company has *Updated* (as opposed to *Redoing*) the PHA over several revalidation cycles. Document retention and revalidation logistics (e.g., determining what was reviewed, what is still valid) become increasingly complex as more and more MOC documentation is appended to the PHA report. *Redoing* the PHA is a way to simplify the PHA documentation before it becomes unmanageable. Refer to Chapter 8 for more guidance on documentation issues.

Experience shows that a well-organized and conducted *Update* can effectively address the changes that occur over a revalidation cycle. However, some companies require a *Redo* of the PHA after a specific number of cycles (e.g., every second or third revalidation), as discussed in Section 3.3.6. This is done for several reasons:

- Sometimes overlooked errors in a PHA will perpetuate from one *Update* to the next.
- Changes may have been missed in a previous *Update* and a new team, unbiased by previous reviews, is more likely to capture these changes for review.
- Nuances of risk understanding, tolerance, and general risk practices can change within a company over time, and a *Redo* allows the PHA to be recalibrated for consistency with those evolutionary changes.
- Team members should carefully think through the entire range of potential process upsets and judge the current risk of loss scenarios as an educational benefit to the entire team.

5.1.3   *Combining Update and Redo in a Revalidation*

In most cases, the evaluation of the prior PHA and events since it was conducted identifies a range of issues that should be remedied during the revalidation. At one end of the spectrum, as illustrated in Figure 5-2, the deficiencies may be beyond repair (e.g., an inappropriate core methodology was used) or an *Update* would take more time and effort than starting over, and a *Redo* should be performed. At the other end of the spectrum, the prior PHA was excellent and the few changes that have occurred can be easily incorporated with an *Update*. Many revalidations fall somewhere between those two extremes, with some deficiencies that can be most efficiently repaired by using the *Redo* approach while the balance of the PHA is *Updated*. Thus, in practice, two approaches are discretely applied in varying proportions; they are combined, not blended. Thus, *Update* is applied to those portions of the PHA that are being preserved or edited (changed), and *Redo* is applied to those portions being added or replaced (done again from the beginning).

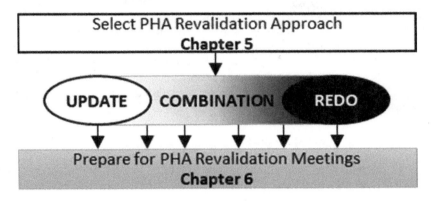

Figure 5-2 Combining Update and Redo in the Same Revalidation

Regardless of the approach or approaches used to revalidate the PHA, the goal of the effort is to ensure a complete, up-to-date, and thoroughly documented PHA. Correcting gaps and deficiencies and revising for changes/incidents creates a valid foundation upon which later revalidations can build. Thus, the decision as to whether to *Redo* just particular portions or the entire PHA would be based, in part, upon how pervasive the gaps or deficiencies were throughout the report.

Several examples demonstrate how *Update* and *Redo* approaches can be used within the same revalidation:

Example 1 – Changes to Risk Ranking System:

Since the prior PHA, the company risk ranking system has changed and now requires the use of a different risk matrix. Otherwise, the *Update* approach would be appropriate. The revalidation team could choose to delete all risk rankings in the existing HAZOP prior to the meeting but leave everything else. The revalidation team would then (1) *Update* the PHA for changes and incidents as appropriate and (2) *Redo* the LOPA calculations of all risk rankings in the HAZOP that still warrant a LOPA.

Example 2 – Required Elements Missing from the PHA:

The prior PHA, conducted using the HAZOP methodology, was reviewed in accordance with the guidance given in Chapter 4, and it was determined that the PHA did not provide a "qualitative evaluation of a range of the possible safety and health effects of failure of controls" as required by a specific company's PHA procedure. The revalidation team could review the HAZOP worksheets in detail and add (*Redo*) consequence rankings (or other qualitative descriptors for the range of effects) to the existing scenarios as appropriate. Remaining changes and incidents would be *Updated* as appropriate.

Example 3 – New LOPA Requirements:

Since the prior PHA, a company now requires a LOPA to be performed on all scenarios that have the potential for a fatality. If the PHA is a suitable candidate based on the discussions in Chapters 2, 3, and 4, an *Update* can be performed on the prior PHA (since it is of good quality, changes have been well managed, incidents are minimal, etc.). Then a new, supplemental LOPA can be performed on the required scenarios extracted from the *Updated* core HAZOP. Note that the term *Redo*, as used herein, includes any analysis being done from first principles, such as the LOPA in this example.

Example 4 – Hazards Were Previously Overlooked:

Assume that hydrogen sulfide (H_2S) release scenarios are both credible and significant in the facility under consideration, and they were not addressed or documented in the prior PHA. If the H_2S was present in only a relatively small portion of the process, the team could choose to *Redo* the affected nodes/sections and *Update* the rest of the PHA.

Example 5 – Issues Associated with Only One Part of the Process:

A PHA of a unit consists of three major process operations (Reaction, Separation, and Compression). Since the prior PHA, several major incidents and changes have occurred, but only in the Reaction section of the process. The revalidation approach could be to *Redo* the PHA for all nodes associated with Reaction but *Update* the Separation and Compression portions.

Example 6 – Batch Processes:

Since the prior PHA, a batch chemical plant had implemented a few, simple changes to the batch process equipment. It had also reformulated one product line in its multi-purpose facility, which required rewriting its batch procedure. The revalidation approach could be to *Redo* the PHA for the re-written recipe, but to *Update* those recipes and equipment with only minor changes.

Example 7 – New Safeguard Requirements:

Since the prior PHA, a company now requires the safeguards credited in the PHA to be included in the facility mechanical integrity program. If safeguards (e.g., the relief valve on a supplier's railcar) are not in the ITPM program, the PHA team is instructed to (1) find or recommend alternative safeguards, (2) make recommendations to include the previously identified safeguards in the maintenance program, or (3) verify that safeguards that cannot be included in the site maintenance program (e.g., due to ownership of the asset by a third party) are being properly maintained and that the ITPM records are auditable. Since every documented safeguard and risk ranking in the PHA HAZOP worksheets will be re-assessed, the revalidation could perform an *Update* for changes/incidents and a *Redo* of safeguards and risk rankings, without altering causes and consequences.

Example 8 – Misapplication of Supplemental Risk Assessments (e.g., LOPA):

In preparation for the PHA revalidation, it was discovered that the associated supplemental risk assessment (e.g., LOPA) was far too liberal in the application of some criteria (e.g., conditional modifiers were overused). This misapplication alone would not be reason to *Redo* the core PHA (i.e., the HAZOP worksheets); however, the team should consider whether (1) only a few affected LOPAs can be *Updated*, or (2) the issue is systemic and all the LOPAs should be *Redone*.

5.2   SELECTING THE REVALIDATION OPTIONS

To determine the approach prior to beginning preparation for the revalidation meeting, it is necessary to carefully review and integrate the information gathered and questions answered in previous chapters of this book. This activity should begin several months before the revalidation due date, which may be immovable due to regulatory requirements. The time required for revalidation preparation, analysis sessions, and document preparation depends upon the revalidation approach, and the revalidation approach depends upon an assessment of the unit's prior PHA and operational history. Working back from the delivery deadline and allowing for contingencies, it is prudent to start the activities described in Chapters 2, 3, and 4 at least two months before the delivery date for a small/simple revalidation, four months for a typical revalidation, and six months for a large/complex revalidation.

5.2.1   *Have the Requirements Changed Significantly?*

As discussed in detail in Chapter 2, external and internal requirements can change over time and impact the revalidation effort. These requirements could include changes such as:

- New regulatory requirements (at a local, state, or national level)
- Changes in company policies and procedures, or loss prevention goals (including changes to supplemental risk assessment procedures (e.g., LOPA, QRA)
- New or revised risk tolerance criteria
- New or revised RAGAGEP or industry standards
- New insurer requirements
- Company merger or acquisition

If the rules the prior PHA followed no longer apply to the current revalidation, then the foundation of the prior PHA should be questioned. Can these issues/changes be fixed in a simple and straightforward manner (*Update*)? If they cannot, then a *Redo* of the PHA should be undertaken.

When considering the scope of a *Redo* initiated because of changes to requirements, the revalidation team should consider how significant these changes are before setting the prior PHA aside completely and starting over from blank worksheets. Significant effort, knowledge, and process understanding exist within the prior PHA.

If the number or significance of changes in the requirements is minor, it may be possible to *Redo* just those portions of the prior PHA to address them, and then proceed to *Update* the rest of the PHA. If there have been multiple, or significant changes, it may be more appropriate to *Redo* the entirety of the PHA.

Changes in supplementary risk assessment requirements can often be addressed using either the *Update* or *Redo* approaches, and sometimes both as these examples illustrate.

Example 1 – New Supplemental Risk Assessment (e.g., LOPA, QRA) Requirements:

Consider Example 3 in Section 5.1.3 where a company has decided to perform LOPA on scenarios with potentially fatal consequences. An *Update* would suffice for the prior PHA. However, the new LOPA requirement effectively changes the rules. LOPA requires that each cause-consequence pair be evaluated separately. Thus, the prior PHA may have to be *Redone* to associate specific consequences and specific safeguards with each individual cause (if they were not previously documented that way). If an *Update* is being performed, the LOPA team will need to be particularly attentive as to whether the PHA safeguards qualify as independent protection layers since the prior PHA team may not have performed the PHA with LOPA in mind.

Example 2 – Changing Supplemental Risk Assessment (e.g., LOPA, QRA) Requirements:

Company A's risk assessment procedures previously required only LOPA to be performed on selected scenarios. New requirements now require a QRA to be performed on high-severity scenarios and LOPA to be performed on all other scenarios. The supplemental risk assessment applied to this PHA will be significantly different after the revalidation. QRA is being added where it was not performed before, and many LOPAs are being added to lower severity scenarios. It should not be necessary to *Redo* scenarios developed in the core HAZOP table simply due to the LOPA changes. The QRA is a *Redo* because it is being performed initially from first principles. The LOPAs could be completely *Redone*, or the team could review and *Update* the existing LOPAs and add new LOPAs for all the other scenarios per the changed requirements.

If the revalidation requirements have not changed significantly, the next step is to evaluate the prior PHA.

5.2.2   Is the Prior PHA Deficient or Unacceptable?

As discussed in detail in Chapter 3, the prior PHA is the foundation on which a revalidation is performed. A questionable foundation may be repairable (by *Update* alone or a combination of *Update* and *Redo*), but each deficiency in need of repair increases the time and effort the team will need to spend to ensure the current PHA revalidation is accurate. The three questions posed in Chapter 3 to judge acceptability of the prior PHA are.

1. Does the prior PHA meet essential criteria?
2. Is the prior PHA of good quality and complete?
3. Are there other reasons to *Redo* the PHA?

Before the revalidation, the prior PHA should be evaluated. Appendix A includes a sample checklist that can be used to determine if the essential criteria are satisfied, and Appendix B includes a checklist that can be used to evaluate the quality and completeness of the prior PHA. The PHA leader or revalidation team will need to consider the number and magnitude of issues related to the prior PHA. This consideration should include an evaluation of the prior PHA core methodology (e.g., What-If, HAZOP), any required complementary analyses (e.g., facility siting study, human factors checklist, damage mechanism review), and supplemental risk assessments (e.g., LOPA, QRA, bow tie analysis).

> **Significant Errors and Oversights in the Prior PHA Affect the Current Revalidation Approach**
>
> Consider Example 4 in Section 5.1.3, where the hazards associated with release of H_2S were not adequately considered in the prior PHA. If the H_2S was present throughout the process, the team would likely choose to *Redo* the entire PHA rather than attempt to repair the entire document. If the revalidation were to be a focused *Update*, and not a more detailed review, the existing issue could propagate into the revalidation without being fully addressed.

If the prior PHA is deficient or unacceptable such that issues cannot be repaired, or if the issues will take more time to repair than starting over, a *Redo* is the better approach.

If the prior PHA is deficient or unacceptable, but those issues are limited in number, minor, and/or repairable, a *Redo* of those issues followed by *Update* is likely the most efficient course of action.

If there are no significant issues related to the quality or completeness of the prior PHA, the next step is to evaluate the operating experience in the interim since the prior PHA was conducted.

5.2.3 *Are There Too Many Changes or Significant Revelations in Operating Experience?*

As discussed in detail in Section 4.2, operating experience and changes since the prior PHA can significantly affect the revalidation approach. The types of operating experience to consider include:

- MOCs and PSSRs (changes)
- Incidents
- Routine maintenance records
- Audit results
- Organizational changes
- Metrics and overall performance

The frequency and magnitude of changes and incidents and the degree to which the changes and incidents have been properly controlled are factors to be considered in determining the appropriate revalidation approach. Answers to the following questions will provide insight as to the completeness of the MOC record:

- How many recommendations in the prior PHA were accepted by management? Is there an MOC for each of them?
- How many revisions have been made in the piping and instrumentation diagrams (P&IDs) for the unit since the prior PHA? A single MOC might involve more than one drawing, and some drawing revisions may simply be text or symbology corrections unrelated to any process change. Nevertheless, there is typically some correlation between the number of MOCs and the number of drawing changes.
- How many new capital purchase items have been made for the unit since the prior PHA? (Be sure to include purchases by small and/or large project budgets, as well as the unit's operating budget for items that were not replacements-in-kind.) A single MOC might involve multiple capital purchases, but there is typically some overall correlation between the number of MOCs and the number of capital purchases.

- Are there any significant gaps in the MOC dates and P&ID revisions since the prior PHA? Does the operating history provide reasonable explanations for such gaps? Unexplained gaps or drawing revisions during MOC gaps often indicate breakdowns in the MOC program and unreviewed changes.

- How many recommendations from incident investigations since the prior PHA were accepted by management? Is there an MOC for each of them?

Numerous or significant incidents, changes (especially uncontrolled changes), maintenance recommendations, or audit recommendations may warrant a *Redo* of the PHA. If the number of changes or incidents has been few, but some of them have been significant, it may be possible to address this by a *Redo* of affected sections and *Update* of the remainder.

Simply stated, it usually takes more time and effort to *Redo* the PHA than to *Update* it. Thus, the *Update* approach is favored for a complete, well-executed, and thoroughly documented PHA with relatively few major changes. Otherwise, if it will take less time and effort to *Redo* the PHA (or portions of the PHA), the *Redo* approach is often favored.

> **Implementing a More Rigorous Complementary Analysis**
>
> After the prior PHA, a company might decide to implement a more rigorous evaluation of human factors and critical task evaluation than in the past. If all or most scenarios in the PHA are to be reviewed, the best approach may be to *Redo* the PHA. However, if the company is just using a more comprehensive human factors checklist, then the checklist could be *Redone* and the rest of the PHA could be *Updated*. See Section 3.3.2 for more guidance on evaluating complementary analyses.

> **Significant Incidents Since the Prior PHA Affect the Revalidation Approach**
>
> Consider Example 5 in Section 5.1.3. The recurrence of major incidents in the Reaction section suggests that there may be issues with the thoroughness of incident investigations and/or the effectiveness of corrective actions. If the revalidation of this PHA is a focused *Update*, insufficient rigor might be applied to the PHA revalidation, and incidents might continue to occur in this portion of the process.

5.3 PRINCIPLES FOR SUCCESSFUL REVALIDATION APPROACH SELECTION

Actual experience in conducting revalidations, by PHA practitioners across a number of companies, has highlighted some keys for success, as well as some things that can impede success. While none of the items listed are absolute rules, they do provide valuable guidance.

Successful Practices:

- Beginning the revalidation approach decision process well in advance of the revalidation due date so time constraints do not unduly affect the choice of revalidation approaches
- Using the *Redo* approach for any new and separate requirements (e.g., a quantitative analysis of facility siting issues) that do not affect the core analysis scenarios, and *Updating* the core analysis
- Reviewing process safety management system audit (internal or external) recommendations and other process safety performance indicators prior to deciding if a *Redo* or *Update* is appropriate, and ensuring any deficiencies are addressed during the revalidation
- Performing a *Redo* periodically (e.g., every 2nd or 3rd cycle)
- If performing a *Redo* based on schedule (e.g., a *Redo* is being performed on the third cycle or 15 years after the initial PHA), including a post-*Redo* gap analysis to help ensure no scenarios in the prior PHA were missed or interpreted incorrectly

Obstacles to Success:

- Completely ignoring the prior PHA when using the *Redo* approach and losing process history and design knowledge contained therein
- Failing to methodically evaluate the prior PHA and operating history when selecting a revalidation approach

6 PREPARING FOR PHA REVALIDATION MEETINGS

This chapter assumes that the revalidation leader has reviewed the prior PHA and recent operating experience as described in Chapters 3 and 4, and then decided upon a revalidation approach as described in Chapter 5. Given that the majority of revalidations use a combination of both the *Redo* and *Update* approaches, Figure 6-1 shows two workflows – one for those portions of the prior PHA being *Updated* and the other for those portions being *Redone*. This chapter discusses preparations important to the success of the overall revalidation effort. The tasks described herein are not the only possible path to success; however, they describe an approach distilled from successful revalidation studies.

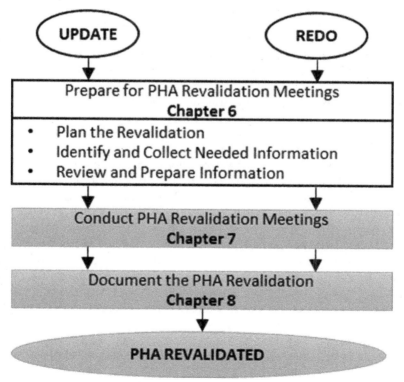

Figure 6-1 Revalidation Flowchart – Preparing for Meetings

6.1   PLANNING THE REVALIDATION MEETINGS

Proper planning of the revalidation meetings is essential for a cost-, time-, and resource-effective effort that will lead to a high-quality revalidation. Most of the tasks are analogous to those necessary for an initial PHA as described in the CCPS book *Guidelines for Hazard Evaluation Procedures* [2]. The planning may be accomplished by the study leader, with the support of other involved parties such as process experts, production management, or the site process safety or PHA coordinator. Common issues to be considered when planning are discussed in this section.

6.1.1   *Establishing the Revalidation Scope*

As described in Sections 1.2 and 1.3, a PHA may encompass complementary and supplemental analyses in support of the core PHA methodology. As discussed in Section 3.1.3, the physical, operational, and analytical scope of the prior PHA should have been evaluated with respect to current requirements as explained in Chapter 2. Based on that review, a revalidation approach should be selected for each element of the PHA as discussed in Chapter 5. Part of that review is to determine whether any complementary or supplemental analyses (e.g., LOPA, bow tie analysis, facility siting study) are to be included in the scope of this revalidation effort, or whether they will be revalidated separately.

Many facilities will prepare a specific scope document/table (sometimes called the *charter, charge, directive,* or *terms of reference*) for the upcoming revalidation effort. This document identifies each major element of the PHA and the plan for its revalidation. It is usually approved formally by management because it effectively commits the necessary organizational resources to the revalidation activities. To establish work priorities, it is best if the scope approval also includes a specific project schedule and the commitment of specific team members.

The scope of the revalidation need not be dictated by the scope of the prior PHA. It is conceivable that an organization might choose to aggregate (or perhaps further subdivide) processes prior to the revalidation effort. For example, it might be decided to combine two PHAs (one nearly five years old and one only three years old) into a single revalidated PHA. There is no apparent impediment to doing so, as long as no single PHA goes beyond its revalidation deadline. In such circumstances, the revalidation team would need to gather information on both PHAs in order to plan the revalidation effort properly.

Table 6-1 is an example of scope topics for a particular PHA revalidation, including specific tasks for the team to complete, *Updating* some and *Redoing* others:

Table 6-1 Example Scope for a Specific PHA Revalidation

PHA Item *(and Method)*	PHA Revalidation Task(s) for Example Unit
HAZOP *(both Update and Redo)*	• MOCs/Changes – *Update* existing HAZOP • Incidents – *Update* existing HAZOP • Previous PHA recommendation resolution – *Update* existing HAZOP • Risk judgments for new risk corporate matrix – Team will re-assess (*Redo*) all risk rankings in the PHA • Identify any new scenarios that should undergo LOPA
Facility Siting Checklist *(Update)**	• Review each question in the previous checklist and validate (or *Update*) response
Human Factors Checklist *(Redo)**	• Review each question and develop a new (*Redo*) response; team will use the new corporate checklist and disregard the previous PHA checklist
Additional Risk Analysis *(Update)*	• Review and *Update* external events checklist • Review any new or affected high-severity fire, explosion, or toxic release scenarios; if revisions are needed (e.g., to the plant facility siting study due to new release points or hazardous inventory)
LOPA *(Update)*	• MOCs/Changes – *Update* existing LOPA • LOPA Recommendation closure – *Update* existing LOPA • New independent protection layer failure rate data is available; ensure the previously claimed PFDs are still accurate and *Update* if necessary • Add new LOPAs scenarios identified during the HAZOP review

* Note: In the example in Table 6-1, the revalidation team has decided to Update the Facility Siting Checklist but Redo the Human Factors Checklist. It is not typical for the same revalidation to Update one checklist and Redo another, but it is certainly possible. For example, in this case, the facility may have made significant progress in its human factors programs and therefore wanted to perform this checklist from the beginning (Redo) to gauge progress without influence from the previous checklist responses.

Multiple PHAs for a Single Process

If multiple PHAs exist for a single process that was started up in stages, the PHA revalidation offers the opportunity to combine these PHAs and ensure no scenarios were missed.

6.1.2   Selecting Team Members

The same issues and considerations discussed in Section 3.1.2 regarding the qualifications of the prior PHA team are equally relevant to the revalidation team. At a minimum, the revalidation team must have the same set of skills and qualifications required by local regulations for any PHA team. Globally, the minimum required skills usually include:

- Engineering expertise
- Operations expertise
- Expertise in the analysis method being used (e.g., HAZOP)

Beyond that, depending upon the specific scope of the PHA, there may be requirements (See Section 2.2.) for additional team skills, such as:

- Maintenance expertise
- Instrumentation and controls expertise
- Process chemistry expertise
- Human factors expertise
- Expertise in risk analysis (e.g., LOPA)

The team composition will likely have to be modified slightly when the complementary analyses are performed, particularly the facility siting checklists. Team members with knowledge of the emergency response plan and its execution, communication systems, and other non-process knowledge will be needed, unless the team that has been studying the other nodes have that knowledge. Another consideration is that in addition to the required skills, there may be other required qualifications. For example, engineering expertise may have to be provided by degreed engineers in particular disciplines or with particular licenses. Operations expertise may have to be provided by hourly employees or field supervisors.

Two options for selection of PHA revalidation team members include (1) for efficiency, keep the same team as before, or (2) for objectivity, select a new team. Either approach can be defended on its merit. If the *Update* approach is selected, the same team may be preferred, but organizational issues (e.g., loss or reassignment of past team members) often prevail, and the team ends up as a mix of new personnel and

> **Team Members for Redo**
>
> If the PHA is being *Redone* because of quality or complete-ness issues, it is usually better to select new team members with a different study leader.

personnel who were involved in the prior PHA. If the *Redo* approach is selected, a substantially or completely different team is more commonly used and sometimes required. In addition, review of the prior PHA may indicate the need to supplement the revalidation team with members having specific areas of expertise not present on the prior team (e.g., metallurgist, control engineer, chemist, loader/unloader). Some of these team members may be part-time or on-call members.

As is the case with any initial PHA, the study leader should make every effort to ensure that the proposed revalidation team has a proper mix of knowledge, experience, and training. For the revalidation to be effective, participants should have specific knowledge of the process being evaluated in addition to general subject matter knowledge. Thus, most organizations have individual or collective experience requirements. For example, at least one of the individuals providing operations expertise might be required to have been a qualified front-line operator for at least five years and to have worked in the subject unit for at least two years. Alternatively, collectively among the revalidation team participants, the organization might require at least three years of operating experience on the subject unit.

Some organizations also place maximum requirements on individual job roles or experience. The concern is that an individual with decades of experience or a senior management position might dominate the team discussions, particularly if others have minimal or no experience with the subject process. Thus, they write their team composition rules to favor teams with a range of experience, believing that inherently leads to more diverse opinions and better brainstorming during the revalidation. Similarly, some organizations limit the number of previous PHA participants on the revalidation team. While it is very beneficial to have some previous participants on the team, there is also a huge benefit to having some new team members who are more likely to raise new issues or challenge previous judgments. When preparing for the revalidation, the study leader should solicit appropriate individuals to fulfill the skill and experience requirements, and they must commit their time to participate. Any identified deficiencies or other concerns regarding the staffing of the team should be resolved with the responsible member(s) of management.

Another consideration is selection of the PHA revalidation leader. The leader drives the study to ensure competent application of the analysis method in conformance with the revalidation approach, and compliance with regulations. If the previous PHA was found to be deficient in those aspects, site management should select a different PHA facilitator with a fresh perspective and skills that are more appropriate.

Once the team composition is established, the study leader should assess what training those individuals need for effective team participation and determine the most effective method to provide that training, as discussed in Section 7.2.2. Given the other constraints (e.g., normal work locations, calendar deadlines, team meeting hours, project budget), it may be difficult to identify the most efficient way to provide the necessary training.

The study leader should first decide what training requirements are specific to the individuals versus those that address general needs of the team. For example, everyone needs a general understanding of process operations and the facility layout. If most team members already have at least a basic familiarity with the process and equipment, then the individuals who need a tour and/or process review can schedule it at their convenience before the revalidation begins. On the other hand, if several participants need such training, then it will be more efficient to begin the first meeting with a review of the process and a brief tour of the facility. The revalidation meeting dynamics are further discussed in Chapter 7.

6.1.3   *Estimating Schedule, Time, and Resources*

During review of the prior PHA and recent operating experience, the revalidation leader should develop good insight into the resources that will be required for the revalidation. If the *Redo* approach is chosen, the revalidation schedule should approximate the schedule of the initial PHA for a new unit of similar size and complexity. Note that this may be significantly longer than the original PHA of the unit because of evolutionary changes in the unit itself (MOCs) or new PHA requirements (e.g., LOPA or damage mechanism review). If the *Redo* uses some salvageable content from the prior PHA as the starting point (rather than completely blank worksheets), the overall revalidation effort should be reduced.

Estimating the time required for a revalidation using the *Update* approach is much more difficult. Revalidation of a simple system with relatively few changes will still require a significant fraction of the original PHA effort, even if the team chooses to read the prior document and reaffirm its accuracy. Units with more numerous and/or complex changes may require as much (or more) time as the original effort. If some portion of the prior PHA must be *Redone* to repair specific defects, additional time and resources will be required. All the issues discussed in Chapters 2, 3, and 4 should be considered, such as:

- The number and complexity of new PHA requirements
- The quality and completeness of the prior PHA
- The quality and completeness of the PSI
- The rigor of MOC/PSSR documentation
- The number of undocumented changes
- The learnings from operational experience

Any estimates of revalidation meeting time usually assume that all the modifications to the process (and any associated documentation) will be reviewed prior to the meetings by the revalidation leader or other experienced personnel. This ensures that personnel who are very familiar with the modifications can explain the rationale for each change and facilitate evaluation by the full revalidation team.

The best strategy for scheduling is to begin with the end in mind. In some jurisdictions, authorities may impose severe penalties for missing revalidation deadlines. When stating the allowed interval between PHA revalidations, most regulations and company policies refer to "completion" of the prior PHA, but some do not clearly define that term.

When is a PHA complete? Many companies define the completion date as the calendar date on which a final authorizing signature is affixed to an *Approvals*, *Authorization*, or *Acceptance* page of the report. Other definitions of the completion date include: (1) the last meeting day of the core PHA team; (2) the day the PHA facilitator assembled the complete PHA report (including core, complementary, and supplemental analyses); (3) the day the PHA report and/or its findings were transmitted or presented to management; or (4) the day management formally accepted the report, usually accompanied by management's initial response to the PHA team recommendations. The definition of PHA completion should be clearly stated in the organization's PHA policy and guidance documents.

Regulations generally leave the precise definition of a revalidation completion date to the discretion of each organization, as long as it is being applied consistently and accomplishes the intent of regular PHA revalidations within the specified cycle. An additional expectation is that management will act in a reasonable time on PHA team recommendations.

As a practical matter for the PHA facilitator, the goal is to get the revalidation completed and the results presented to management, so the official completion can occur and the PHA cycle (often set at every five years) be maintained. The PHA revalidation must be started far enough ahead of the required completion date to allow a high-quality product to be produced in compliance with any internal company policy or external regulatory requirements. One approach is to schedule the revalidation meetings to begin five years after the first meeting for the prior PHA, since most revalidations can be completed in less time than the original PHA.

The preparation work described in Chapters 3 through 5 provides a definition of the expected scope of the revalidation effort and a basis for estimating a reasonable schedule for the PHA revalidation meetings. If the *Redo* approach is selected, it is prudent to schedule time for the revalidation as if it were the initial PHA of the unit. That automatically provides some contingency in the planned schedule. Then the proposed schedule can be further adjusted to help ensure the availability of required team personnel and address other issues (such as an

> **Does a Kickoff Meeting Satisfy PHA Revalidation Completion?**
>
> Not typically. Caution should be used if defining the meeting start date as the new PHA cycle. Due to factors such as team member availability, unit upsets, and turnarounds, there may be significant delays before completion of the revalidation meetings. If a revalidation is kicked off, but the bulk of analysis and recommendation development do not occur for many months (or years) it is very likely the required cycle for revalidation completion will not be achieved.

upcoming unit outage or the start of holiday season). If the *Update* approach is selected, the typical expectation is that the revalidation can be accomplished in less time. However, the more an *Update* tends toward the *Redo* end of the spectrum, the more time is typically needed for meetings and report writing, resulting in the need for an earlier start date relative to the required completion date. Careful planning and negotiation will help ensure that the planned schedule can be successfully executed with the available resources.

6.2 IDENTIFYING AND COLLECTING INFORMATION

As discussed in Chapters 3 and 4, considerable information should have been identified, collected, and evaluated to decide which approach will best meet the revalidation goals. This information needs to be organized and available for use during the meetings.

Paper vs. Electronic Information

The revalidation team must have access to any information necessary. Whether this information is available as a printout in the PHA room or in electronic form is usually a PHA team preference. Each team member should be provided a copy of the current P&IDs with analysis nodes identified (usually highlighted in various colors). To the extent that other information is digitized, the team leader may want to compile a much larger electronic library or roadmap of relevant information easily accessible to all the team members.

6.2.1 Determining Information Requirements

Table 6-2 lists several information resources that can be used in a PHA revalidation as well as comments on the purpose and usefulness of the information. Some information needs are particular to the analysis method(s) that will be used, as discussed in the CCPS book *Guidelines for Hazard Evaluation Procedures* [2]. Regardless of the approach selected (*Redo* or *Update*), current and accurate information is critical to the success of the PHA revalidation.

An index of other information reviewed by the team leader and available to the team will expedite retrieval of that information (e.g., safety relief valve sizing) if a specific question arises during the revalidation meetings.

If the *Redo* approach is selected, the information needed, at a minimum, is the same as for an initial PHA. Typically, additional information is available, such as the operational history, maintenance history, prior PHA results, and audit findings. Current and accurate P&IDs are particularly important if the *Redo* approach was chosen because a different core analysis method or different node definitions will be used. Even if the PHA is being *Redone* for other reasons, the drawings and procedures should show how all the additions and deletions resulting from changes affected the prior node definitions. MOC documentation should not be required for a *Redo* because the revalidation team is making an assessment based on the current configuration of the process. Up-to-date configuration should be embodied in the current process safety information and procedures.

If the *Update* approach is selected, in addition to the information required for a *Redo*, the team will definitely need copies of MOCs for all permanent changes and all currently active temporary changes. The revalidation leader may flag particular deviations or questions that appear to need updating based on their preliminary assessment of each modification. For easy reference, some revalidation leaders attach (or electronically link) a copy of the relevant MOC(s) to the affected nodes or identify the changes to be discussed on the current set of highlighted drawings. This preparation accelerates the pace of the PHA revalidation meetings and allows the PHA revalidation leader to guide the team methodically through the *Update.*

Table 6-2 Example Documentation to Collect for a Revalidation

Information Type	Comments
1. Prior PHA report(s) and related documentation	Part of the baseline for revalidation is established. More than one report may be involved. Include reports for projects and changes. See MOC/PSSR.
2. Resolution completion report(s) for prior PHA recommendations	Part of the baseline for revalidation is established. Recommendations remaining open should be resolved or factored into the revalidation.
3. MOC and PSSR documentation • For processes, procedures, and equipment to be studied • For interconnected processes, with their procedures and equipment, as appropriate • For recurring or prolonged temporary changes • For organizational changes	Identifies the controlled (documented) changes made since last revalidation. Serves as a database against which uncontrolled (undocumented) changes, if any, can be identified. See Section 5.2.3.

Table 6-2 continued

Information Type	Comments
4. Incident reports • For process to be studied • For similar processes at other sites	Weaknesses in the PSM program or prior PHA may be identified. Consider as source of additional scenarios during revalidation. Reports should cover period since last revalidation.
5. Audit reports	Weaknesses in PSM program or prior PHA may be identified. Reports should cover period since last revalidation.
6. Process upset reports and investigations	Information is provided on events that did not trigger the near-miss criteria, but they may still be indicative of a potential process problem.
7. Piping and instrument diagrams • Current, up to date • Those used for prior PHA, if available	P&IDs serve as the "road map" for study sessions. Comparison of current P&IDs with those of prior PHA helps identify potential changes when performing an *Update*.
8. Block flow or process flow diagrams	The processes or facilities under study are described.
9. Area electrical classification drawings	The processes or facilities under study are described.
10. Process chemistry description and interaction matrix	The processes under study and reactive chemical hazards are described.
11. Safe operating limits for the process	The processes or facilities under study are described.
12. Material and energy balances • Current, up to date • Those used for prior PHA, if available	The processes or facilities under study are described.

Table 6-2 continued

Information Type	Comments
13. Descriptions of safety systems and records of their functional testing and actual performance history • Cause and effect diagrams • Alarm reports	Safety systems are commonly relied-upon safeguards. Documentation is required to ensure that claimed safeguards are in place and will function properly when needed.
14. Operating procedures • Current, up to date • Include all operating modes • Those used for prior PHA, if available	Operating procedures serve as a valuable reference, especially for batch processes. They identify steps where specific, required actions are critical. They can also be used to identify non-routine operating modes for analysis, as well as any uncontrolled or undocumented changes.
15. Pressure relief documentation, sizing calculations, and test reports	This documentation can assist in evaluating safeguard capability and reliability.
16. Mechanical integrity information • Maintenance records and procedures • Equipment inspection reports • Equipment design information	This information can assist in evaluating equipment reliability and potential internal/external failure modes (e.g., overheat damage, corrosion).
17. Safety and emergency maintenance work orders	Undocumented changes, frequent causes, or poor safeguard reliability can be identified.
18. Consequence assessments • Siting studies • Dispersion studies • Fire/explosion studies	The team can use the assessment to calibrate; with respect to how serious the effects of a release could be and can be used to assist in consequence/risk ranking.
19. Safety data sheets	The sheets include property data on chemicals used in the process, service systems, etc.

Table 6-2 continued

Information Type	Comments
20. Results from safety culture survey and employee interviews	System weaknesses not detected via PHA or incident investigations may be revealed.
21. Safety committee reports	Operational concerns may be identified.
22. Other pertinent process safety information	Plot plan drawings, critical instrument lists, etc., may be included. New or previously overlooked hazards may be identified. Uncontrolled or undocumented changes may be identified.

6.2.2 Distributing Information

Certain key information items (or a link to the electronic library) can be provided to the team members for their review prior to the revalidation sessions. This helps team members prepare so that they will be more effective during the meetings. If it is practical to archive a copy (or index) of the reference information with the PHA, it will also be useful in future revalidation team preparations. The value of distributing printed copies of information before the meeting should be evaluated on a case-by-case basis.

6.3 REVIEWING AND PREPARING INFORMATION

The leader (or a designee) should also prepare worksheets for recording the meeting notes. These worksheets include the core analysis (e.g., HAZOP, What-If, FMEA), as well as any complementary analysis (e.g., human factors checklist) or supplemental risk assessment (e.g., LOPA).

6.3.1 Prior PHA Reports and Related Documentation

When performing a *Redo*, there should be a worksheet for each node of the analysis (just like the initial PHA), but the exact form will depend on the analysis technique(s) and the documentation format being used. If the leader has decided to use some information from the prior PHA as a brainstorming aid, that information should be made available to the revalidation team. The worksheets used for an *Update* are copies of the worksheets published in the prior PHA, plus copies of the worksheets for any added nodes, minus the worksheets for any removed nodes. Although this sounds easy, it may involve a significant amount of work. The nodes and formatting used in project MOCs may not align with the nodes in the prior PHA, and they rarely show any equipment that was removed. Thus, depending on the number and complexity of the changes, the revalidation leader may want to map the nodes from the prior PHA onto a current set of drawings and then modify, add, or delete nodes as appropriate.

The leader should then generate worksheets corresponding to the revised nodes, particularly if the worksheet formatting and organization is being substantially changed (e.g., large nodes are being separated into smaller nodes because the core analysis method changed). However, if any information (e.g., causes, consequences, and safeguards) is transferred from the prior PHA into the new worksheets, the revalidation team should be alert to errors and omissions as they *Redo* each node. However, if the node revisions were simply necessary to include new equipment and the MOC hazard reviews were conducted with the future PHA in mind, the leader can directly transfer information from the MOCs. The resulting PHA tables (e.g., HAZOP worksheets, checklist responses) look very similar to those in the prior PHA and can be evaluated using the *Update* approach. In either case, if the re-organization creates blanks or adds items the revalidation team will have to consider, the worksheet fields can be highlighted or flagged for searchability.

6.3.2   *Prior PHA Recommendation Resolution Status*

If a *Redo* is being performed, the status of prior recommendations should already be reflected in the current documentation (PSI, procedures, etc.) available for analysis. However, having a list or summary of the prior recommendations and their resolutions is the easiest way for the revalidation team to ensure those issues were adequately addressed and is useful data for comparison with the recommendations from the *Redone* PHA.

If an *Update* is being performed, the resolutions of prior PHA recommendations are critical inputs to the *Update*. During the PHA revalidation sessions, the team will address each previous recommendation by updating the affected scenarios. Table 6-3 is an example of how one might document prior PHA recommendations in a revalidation. Note that in this table, the Rec. No. and Resolution columns are completed during preparation, and the PHA Revalidation Team Comment should be filled out during the meetings with the team.

MOCs for Completed PHA Recommendations

Implementation of a prior PHA recommendation should involve an MOC with associated hazard review. Identifying the associated MOC number for each such recommendation prior to the meetings will expedite the meeting by having this information more readily accessible to the revalidation team.

If recommendations from the prior PHA were implemented without a documented MOC, the hazards associated with the changes will need to be considered during the current revalidation meetings. The fact that any changes were implemented without an MOC might initiate a discussion between the facilitator, the revalidation team, and management prior to start of the meetings. Is a *Redo* warranted? Are there other issues?

Table 6-3 Example of Prior PHA Recommendations with Comments

No.	Resolution	PHA Revalidation Team Comment
1	*Status 1 detail*	The team inserted new Node 12 containing the equipment added from this recommendation.
2	*Status 2 detail*	The team updated in safeguards in Node 1, Deviation 1.6 per completion of this recommendation. See MOC 1 for details.
3	*Status 3 detail*	This recommendation was open and in progress. It should be completed at the next opportunity.
4	*Status 4 detail*	This recommendation was closed shortly after the prior PHA without action. The team concluded that it should be completed and made a new recommendation in Node 3, Deviation 3.1 to address this concern.
5	*Status 5 detail*	This recommendation was closed shortly after the prior PHA without action. The team agreed and removed it from Node 4, Deviation 4.3.

6.3.3   MOC and PSSR Records

As previously noted, one of the basic reasons for PHA revalidation is to address changes that have occurred since the prior PHA was conducted. Consequently, one of the most important, and time-consuming, preparatory steps for the revalidation effort is the identification of those various changes. A full understanding of the number and nature of the changes is key to determining the scope of the revalidation effort, and to completing an effective revalidation.

If a *Redo* is being performed, the intent of the *Redo* is to conduct the PHA as the process is currently configured with up-to-date PSI and procedures. Thus, in theory, the MOC and PSSR documentation should not be needed because the changes made to the process since the prior PHA should be reflected in the current information the revalidation team will analyze and/or reference. However, if the robustness of the MOC program is questionable since the prior PHA, the information on MOCs and PSSRs for permanent changes could be collected. The information could then be examined (or audited) for any discrepancies between the "as found" condition of the process and the P&IDs. The review to ensure that all relevant updates to P&IDs and/or operating procedures are properly considered could be conducted as a separate activity in preparation for (or during) the revalidation meetings themselves.

If an *Update* is being performed, the MOCs and PSSRs since the prior PHA are critical inputs to the *Update*. During the PHA revalidation sessions, the team will address each change by updating the affected scenarios. Change documentation is often lengthy and can contain more information than is possible to present to the PHA team in a concise and helpful manner. During preparation, this information should be gathered and organized in a table that can be reviewed with the team, with notation of the location where more detail can be found. (See Appendix C.)

Ideally, all changes would be carefully controlled and implemented via the MOC and PSSR systems, and the changes recorded in those systems would be easily collected and logged; however, some changes may have slipped through the system. It is important that changes, whether or not they are controlled or previously documented, be identified for consideration during the revalidation effort. In practice, it is extremely difficult for a revalidation team using the *Update* approach to compensate for a poor MOC system, despite their best efforts.

Table 6-4 lists examples of the types of changes that may be encountered. Experience has shown that there are a number of productive sources of this change information. These will be discussed in detail in this section, along with suggested strategies for identifying the changes. While these suggestions will apply to many situations, documentation practices vary from one company to the next, and alternative strategies may produce superior results in some situations.

Table 6-4 Example Types of Changes to Consider in the Meetings

Process/System Characteristic	Examples of Change
Hardware	Equipment added/removed/re-configured Piping tie-ins Relief systems Utility tie-ins Vessels re-rated
Safety Systems	Safety shutdowns added/removed or new trip points Safety dumps or purges added, removed, or modified Fire detection/suppression systems added, removed, or modified
Control Systems	Conventional system replaced with a distributed control system (DCS) Safety programmable logic controller (PLC) added Safety Integrity Level (SIL)-rated system installed Alarm rationalization Alarm setpoint change
Technology	New chemicals used or produced Chemical inventory increases/decreases Changes in the process, including chemistry and new operating modes Changes in safe operating limits Unit throughput increases/decreases
Operations	New or revised procedures Staffing decreased/increased Training program modifications Changes in occupancy patterns/location of personnel work areas Fatigue Risk Management Program modified Safe Work Practices modified
Management Systems	Audit frequencies adjusted Outsourcing engineering work Revised emergency evacuation plans

Table 6-4 continued

Process/System Characteristic	Examples of Change
Siting	Addition, relocation, or removal of temporary structures or trailers
	Addition or demolition of permanent structures
	Changes in occupancy
	Changes to emergency vehicle access routes
	Increased vehicle traffic
	Equipment relocation
	Electrical classification changes
	New adjacent units or structures
	Increases in, or addition of new, vulnerable exposures surrounding the process unit (either on-site or off-site)
	Process changes that introduced new scenarios with potential siting impacts

Identifying Documented and Controlled Changes. Appendix C contains an Example Change Summary Worksheet that could be used as an aid for this task. This worksheet is organized by P&ID number (first column) so that the revalidation team can evaluate each change in the context of the P&ID-based review during the revalidation work sessions. The worksheet provides for documentation of the nature of the change, the source of information regarding the change (i.e., in what documentation was the change discovered), and explanatory comments. The last column of the worksheet in Appendix C allows documentation of the proposed action for addressing the change during the revalidation.

Table 6-5 is another example of how to document changes reviewed in a revalidation. Note that in this table, the MOC Reference and Change Description columns are completed during preparation, and the PHA Revalidation Team Comment should be filled out during the PHA sessions with the team. The MOC Detail column provides the minimum detail required, and the addition of the PHA Revalidation Team Comment column makes the documentation a stronger record of how the *Update* approach was applied. This information would be useful in the PHA revalidation report document.

Table 6-5 Example Table of Changes with Team Comments

MOC Reference	Change Description	PHA Revalidation Team Comment
MOC# 123456	*Change 1 detail*	Because of this change, associated Cause in Node 1, Deviation 1.4 was deleted.
MOC# 123457	*Change 2 detail*	Updated safeguard TI-123 in Node 1, Deviation 1.6 per completion of this change. Associated LOPAs were also updated. See previous PHA Recommendation 2 for further details.
MOC# 123458	*Change 3 detail*	This change had no effect on the PHA, and no updates were made.
MOC# 123459	*Change 4 detail*	New Nodes 31 and 32 were added because of this change. Node 30 was reviewed in detail due to rerouting of process piping. Comments were added in the HAZOP worksheet.
MOC# 123460	*Change 5 detail*	This change affected the reliability of several equipment items. Therefore, Nodes 1-3 were reviewed in detail to ensure risk rankings were appropriate. Comments were made in the worksheets if risk rankings were updated.

There often may be no action required to address a particular change during the revalidation. However, those changes that did not receive adequate review prior to implementation (e.g., did not go through the MOC process or only received a cursory review) or temporary changes that are still in service beyond their authorized service date should be flagged for more careful consideration during the revalidation.

Identifying Undocumented and Uncontrolled Changes. Some changes may not have received a thorough MOC evaluation; for example, sometimes the person responsible simply did not recognize what they were doing as being a process change because it was so small, subtle, or incremental in nature. Additionally, some types of change could have a potential process safety impact, but that are not overtly included in an MOC program focused on process technology, equipment, and facilities. Such changes might include:

- Changes in staffing level or key personnel that might impact a facility's capability to respond, or the speed of response, to a particular abnormal situation

- Substitution of contractor operating or maintenance personnel for company personnel
- Sales of assets within the plant to another company
- Procedural changes (particularly frequencies of performing tasks)
- The addition of new structures (congestion), demolition of nearby facilities, or an increase in the number of people in existing buildings near a process, potentially impacting the results of prior siting reviews

Although these types of changes should be included in the facility MOC program, if they are not, it is still part of the PHA revalidation effort to ensure that the hazards of the process are appropriately assessed and controlled. If a list like Table 6-5 were being used to track changes reviewed in the revalidation, a brief synopsis of any additional changes should be appended to the table.

The Checklist of Process, Facility, and Human Factors Changes in Appendix D, may be useful in helping the PHA revalidation team identify and evaluate changes that may not have been documented under the MOC system. Additionally, several of the more common sources of information used to identify undocumented and uncontrolled changes are discussed.

P&ID Comparison. P&IDs are typically the focal point of the PHA sessions and represent one of the more concentrated sources of information about the process. As such, these drawings are a logical place to search for changes, through comparison of the current P&IDs with those used during the prior PHA (if available).

The simplest way to identify modifications to the P&IDs is to check the revision numbers on the current drawings versus the revision numbers on the drawings used during the prior PHA. Another method for identifying changes using the P&IDs is to compare each old and corresponding new drawing to identify any new equipment or lines. For every revision after the prior PHA, the nature of the change and other relevant information (e.g., an MOC record identifier) should be documented for consideration during the revalidation sessions.

> **P&ID Preparation for Changes Review**
>
> Consider marking changes on the P&IDs subject for review during preparation (before the meetings). This will help identify both controlled and uncontrolled changes as well as reduce time in the meetings required to find each change.

Relevant MOC, PSSR, and/or PHA records should be identified for the changes found on the drawings. If no such documentation can be found, this fact should be documented, and the associated change flagged for more detailed review during the revalidation.

Interviews of Facility Personnel. Interviews may be considered as part of the revalidation process to verify that PSI is accurate, and to identify changes that might not otherwise be identified through PSI reviews (e.g., such interviews may be helpful in validating the content of the P&IDs). For example, the interviews may identify frequent or prolonged usage of automatic control valves in manual mode or may identify frequent alarms that are disregarded as a nuisance.

Changes identified during the interviews should be listed, so that they can later be cross-referenced against the MOC records to determine whether they are documented or undocumented changes.

Maintenance Records. Reviewing each work order or reviewing the maintenance history generated during the five years since the prior PHA would be a daunting task and should not normally be necessary. However, a sampling of such records might confirm whether a more detailed review is warranted. Some computerized maintenance management systems (CMMSs) may be able to conduct key word searches for indicators of the type of maintenance work performed (e.g., "upgrade," "change," or "modify") to help identify work orders for consideration.

If work orders describe activities that are more than just *replacement-in-kind* and no associated MOC documentation can be found, these *changes* should be flagged for more detailed consideration during the revalidation sessions.

Purchase Specifications and Records. Changes in purchase specifications or suppliers for chemicals, spare parts, equipment, etc., associated with the process are also changes for which there may be no formal MOC. These types of records may also warrant an inspection to determine if discussion during the revalidation would be useful.

The goal is to prepare worksheets that will facilitate accurate and efficient revalidation meetings. All changes can be documented in a table such as Table 6-5, with a brief description in the center column or on a supplemental page for those without reference number. If done well, the revalidation team will not need to restructure the worksheets during the meeting as they modify the worksheet content with any changes. For example, during an *Update*, a prior recommendation for a low flow alarm might be deleted from the Recommendation column of a PHA worksheet, but the low flow alarm installed to resolve that recommendation would be added to the list of safeguards against the low flow hazard. A complete set of worksheets (some revised, others unchanged) can then be published in the *Updated* PHA.

6.3.4   *Audit Results*

PSM system audit results should have been included in reviewing the operating experience for the PHA as discussed in Chapter 4. Fundamental deficiencies in the prior PHA could have led to the decision to *Redo* the PHA. If a decision has been made to *Update* the PHA, audit results could provide insight into items to focus on, ensuring they were adequately addressed in the prior PHA.

It should be noted that, typically, PSM system audits are conducted using some sort of sampling scheme to identify documentation that is to be reviewed. Thus, the fact that a particular PHA report is not mentioned in an audit finding should not be construed as conclusive proof of the quality of that particular PHA; the PHA may not have been examined as part of the audit.

6.3.5   *Incident Reports*

Accident and near-miss reports may identify loss scenarios that warrant consideration within the revalidation effort and should be reviewed as part of the preparation. Where relevant information can be gathered on incidents in similar facilities or processes, such information may also be considered. This information can be collected from open literature sources or specific databases such as the CCPS Process Safety Incident Database (PSID).

A major loss, a series of less significant losses, or numerous near-miss incidents in a process unit or similar unit can be indicators of a weakness in, or failure of, some PSM system element(s). Incidents involving the subject process should be scrutinized to see if the particular circumstances prompt a concern with the quality of the prior PHA. For example, the PHA team may not have identified the potential for a particular incident. Alternatively, the PHA team may have identified the potential for the incident, but erroneously judged the safeguards to be adequate.

The incident investigation reports, their recommendations, and the details of their resolution should be reviewed in a manner analogous to that previously described for PHA recommendations or changes. Using a format similar to Table 6-5 will allow the revalidation team to document their review of incidents and results of the review.

> **Example – Evaluating Incidents**
>
> A process experienced an incident where material was released during abnormal conditions and a fire occurred. The prior PHA team did not consider the material to be flammable at normal operating conditions. The prior PHA should be reviewed considering this new information and updated accordingly.

6.3.6   Current Piping and Instrument Diagrams

P&IDs are typically the focal point of the PHA meetings, and performing a revalidation based upon incorrect or out-of-date drawings would be a waste of time and resources. Particular care should be devoted to ensuring that the P&IDs clearly and accurately reflect the process and equipment details needed to support the revalidation effort. It is prudent to spot check some drawings in the field to confirm their accuracy. While the cost of field verifying and updating these drawings can be significant, if the condition of the P&IDs requires such updates, it is imperative that this be accomplished prior to the start of the revalidation.

6.3.7   Current Operating Procedures

It is common for initial PHAs of continuous processes to focus only on normal operations, failing to address non-routine, critical operating modes such as startup, shutdown, preparation for maintenance, emergency operations, emergency shutdown, and other activities with characteristics that may differ considerably from normal operations. Similarly, PHAs of batch processes may neglect operations such as vessel clean-outs, water runs, or pressure testing.

Experience indicates that many accidents do not occur during "normal" operation but, rather, during such non-routine modes of operation. Consequently, it is important that a PHA evaluate the hazards of a process during non-routine as well as normal (routine) operating modes. Careful review of the procedures for all modes of operation can help to identify these non-routine operations. In addition, discussions with revalidation team members may identify other such non-routine operations. Despite any assurances that the written procedures are accurate, it is prudent to spot check a few of the ones that will be reviewed during the revalidation with front-line workers in the field.

6.3.8   *Special Considerations for Complementary Analyses and Supplemental Risk Assessments*

The preparation for revalidation of complementary analyses and supplemental risk assessments is similar to that of the core analysis. However, some additional items to consider during preparation include:

Analysis/Assessment Procedures. Procedures for application of these analyses/assessments can change and may not have been considered when the PHA procedure/requirements were assessed. Ensure the current procedures are being applied. Additionally, because many different techniques/ tools/analyses can be applied and each one might have different requirements, these procedures should be reviewed to determine the scope to be applied in the revalidation.

Subject Matter Experts. The more technical and specific an analysis, the more likely a subject matter expert (SME) will be required during its review. Scheduling the participation of SMEs can be a concern and should be done well in advance of the PHA meetings. For example, industrial hygiene expert during the human factors checklist, process controls engineer during LOPA, and metallurgist during a damage mechanism review.

Timing and Organization. A PHA is comprised of several interrelated analyses, involving various experts. Carefully planning the timing and organization of the various studies will help ensure the overall PHA objectives are achieved with the most efficient use of resources. For example, will supplemental risk assessments (e.g., LOPA) be *Updated* during the core analysis review or afterwards? Because documentation associated with various analyses can be in a different file or location, determine in advance how the changes made to the core analysis will be propagated into the supplemental risk assessments.

Additional Requirements. Some supplemental risk assessments (e.g., LOPA) require validity or verification of independent protection layer performance. If the company has a requirement to verify the performance of IPLs in the LOPA prior to allowing credit, this information will need to be gathered and be available. Complementary analyses may also need additional information. For example, to revalidate the facility siting study properly, the revalidation team may need current data on ground subsidence and surface water runoff to judge the risk of flooding.

6.4   PRINCIPLES FOR SUCCESSFUL REVALIDATION PREPARATION

Actual experience in conducting revalidations, by PHA practitioners across a number of companies, has highlighted some keys for success, as well as some things that can impede success. While none of the items listed are necessary, they do provide valuable guidance.

Successful Practices:

- Defining and approving the scope of the revalidation effort with authority to commit the necessary resources
    - o   If a revalidation team charter has not been issued prior to preparation, consider developing one
- Scheduling the revalidation meetings in advance (e.g., six to nine months) and obtaining the commitment of all required participants to those dates, to allow for preparation and confirmation of team members
- Selecting team members with appropriate expertise, while also considering the need for:
    - o   Experience with analysis techniques
    - o   Consistency with prior PHA efforts
    - o   Fresh insight and objectivity
    - o   Training future participants for PHA revalidations
- Using the information collected to evaluate the prior PHA, evaluate operating experience, and determine the revalidation approach as a starting point for preparation
- Determining the quality of P&IDs in advance, allowing updates to be made prior to the meetings
- Using a checklist for preparation tasks and documentation collection and review
- Defining the time span encompassed by the current revalidation effort such that documentation is not missed
    - o   Information before the freeze date is included in the prior PHA and later information will be included in the next revalidation
- Having major organizational changes to the PHA reviewed by the revalidation team or other knowledgeable person
- Allowing adequate time between the freeze date and the first revalidation meeting date to copy documents and prepare worksheets for the revalidation meetings

- Allowing adequate time between the freeze date and the first revalidation meeting date for team members to review key documents and tour the process, as necessary
- When performing a *Redo,* determining whether some of the prior PHA will be used or referenced by the team instead of completely starting over with blank worksheets
- Changing PHA formatting and organization prior to the PHA (e.g., large nodes are being separated into smaller nodes) during preparation before the PHA sessions

Obstacles to Success:

- Relying exclusively on the revalidation team leader to gather all the information that will be needed by the revalidation team
- Assuming the revalidation meetings will be the top/only priority of essential team members
- Selecting team members based on their schedule availability rather than their technical competence and suitability
- Using revalidation meeting time to perform tasks that could have been completed during the preparation phase (e.g., updating documentation, searching for changes or incidents)
- Changing the documentation style or organization of a PHA during the revalidation sessions when using the *Update* approach. Straightforward changes can be made during preparation; more significant changes may be better handled by a *Redo* approach
- Assuming the MOC hazard reviews were performed with the same level of detail and met all the same requirements as a PHA
- Performing tedious tasks during the meetings, instead of during preparation. Repetitive documentation tasks such as changing "no consequences identified" to "no consequences were identified by the team" can lead to an inattentive/unengaged team
- Changing PHA documentation software without transferring information from the prior PHA before the meeting
- Using less experienced personnel to revalidate with the *Update* approach than the *Redo* approach
- Failing to plan revalidation of the complementary and supplemental analyses to be consistent with changes in the revalidated core analysis

7 CONDUCTING PHA REVALIDATION MEETINGS

7.1 APPLYING ANALYSIS METHODOLOGIES

The facts that drove the selection of a revalidation approach are also important in the technical analyses. Chapter 5 detailed two revalidation approaches (*Redo* and *Update*) and noted that in practical applications, many revalidations involve a combination of both. This section describes how to implement both approaches. This concept is also illustrated in Figure 7-1.

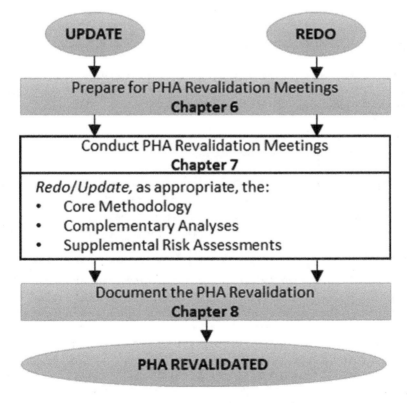

Figure 7-1 Revalidation Flowchart - Conducting Meetings

As discussed in Chapter 1, any PHA can be divided into three parts:

1. **Core Methodology/Core Analysis.** The core of a PHA is the identification of hazards and safeguards, typically using the HAZOP Study, What-If/Checklist, or FMEA technique. This is the foundational analysis that identifies the hazards of the process, the engineered and administrative risk controls (safeguards), and the worst credible consequences if all the safeguards failed.

2. **Complementary Analyses.** The PHA often includes additional analyses focused on specific topics that warrant further consideration using techniques other than those for the core analysis. These complementary analyses are often checklists and include studies such as facility siting studies, human factors analyses, damage mechanism reviews, and dust hazard analyses.

3. **Supplemental Risk Assessments.** Some PHAs include supplemental risk assessments of selected loss scenarios. Often these are Layers of Protection Analyses (LOPAs), but they may include Bow Tie, Fault Tree, Event Tree, or Human Reliability Analyses.

The distinction between these three parts of a PHA is important to the following discussions because the various parts may be revalidated differently, depending on the approach chosen.

7.1.1   Revalidation of the Core Analysis

Redoing the Core Analysis. The *Redo* approach is selected when there are many changes in the process equipment or procedures, the core methodology, existing node definitions or risk tolerance criteria, or when the organization desires an independent assessment of the hazards. If the *Redo* approach was selected, the revalidation is conducted in almost exactly the same manner as the initial PHA. The process is divided into nodes (if the nodes used in the prior PHA are not suitable), and worksheets are prepared. The team identifies potential loss scenarios for each node, and those scenarios with consequences of interest (as defined in the revalidation scope) are documented. At a minimum, the documentation lists the hazards, the existing risk controls (safeguards), and the credible unmitigated consequences if the risk controls were to fail. The documentation usually includes a representative list of specific hazardous events (causes, failure modes) that the team used as the basis for its frequency judgment. The team then decides whether the risk of the scenario exceeds the organization's risk tolerance. If so, the team notes the need for further risk reduction and may offer specific suggestions to achieve tolerable risk.

The main difference between a *Redo* and an *Update* is the starting point. In a "pure" *Redo*, the meeting is conducted in a manner similar to the initial PHA. More often, the team makes some use of the previous PHA documentation if it is of high quality. For example, the previous PHA may be used as a checklist. As the current team concludes its discussion of a particular loss scenario, the team might look at (or display) the previous PHA and ask, "Is there anything we missed?" The differences might be factual (e.g., the previous team included a cause that the current team overlooked) or judgmental (e.g., the previous team took credit for a safeguard that the current team judged ineffective). Regardless, the study leader should continue the team discussion and resolve any discrepancies before moving to the next topic.

Extensive use of the previous PHA is possible as long as it does not compromise the reason the *Redo* approach was selected. For example, perhaps the PHA is being *Redone* simply because the study leader believed it was the more efficient approach, given the large number of changes that affected most of the nodes. With no indication that the previous PHA was deficient, the study leader might pre-populate the revalidation worksheets with some information from the previous PHA and use it to expedite the current discussions as if it were an *Update*. On the other hand, if the *Redo* approach was selected because the organization truly wanted a fresh look at the risks of the entire process, such extensive use of the previous PHA might seriously compromise achievement of that goal. In any case, the revalidation team should be fully involved in developing and analyzing loss scenarios. The facilitator should ensure that any pre-populated entry is discussed, and that additional brainstorming is allowed (and encouraged) beyond those borrowed entries.

Updating the Core Analysis. The *Update* approach is usually selected when there are relatively few, specific changes in the process equipment or procedures, and the core methodology, existing node definitions, and risk tolerance criteria are unchanged. In that case, the revalidation approach is relatively simple and quick to apply, but it still requires thoughtful consideration. The revalidation leader guides the team through the existing documentation, soliciting the team to either confirm or correct the current information. The more detail (e.g., specific valve or instrument numbers) documented in worksheets of the prior PHA report, the easier it is to *Update*. (If the prior PHA lacks such detail, the team should consider adding it to facilitate future revalidations.) When known changes are encountered, the team appropriately edits the affected documentation. Even if there are no known changes, the team should critically evaluate each node for technical accuracy and thoroughness in identifying hazards. The previous team may have made a mistake, a change may have eluded the management of change system, a safeguard relied upon by the previous team may no longer be effective, or incidents may have shed new light on the effectiveness (or ineffectiveness) of safeguards credited in the PHA. The team should also consider the implications of recurring or prolonged operational issues along with factors such as aging, degradation, and obsolescence.

If a combination of the two approaches is allowed, the revalidation team will often *Update* whatever is possible and *Redo* everything else. For example, the revalidation might involve relatively few changes, but the organization's tolerance for risk is different. In that case, the team members should be informed of the reason(s) for, and the contents of, the change record prepared ahead of the sessions. (See Section 4.2.1.) The study leader would then guide the team through the core analysis much like an *Update*, but the risk decisions in every node would need to be *Redone* in light of the organization's revised risk tolerance.

7.1.2   Revalidation of Complementary Analyses

The revalidation approach for complementary analyses is largely independent of the approach chosen for the core analysis. For example, it may be adequate to simply *Update* a facility siting study (qualitative or quantitative), even though the core study was completely *Redone*. Conversely, it may be necessary to *Redo* a human factors analysis while simply *Updating* the core analysis. Some complementary analyses may be performed by experts, apart from the team re-validating the core study. Thus, a damage mechanism review may be performed by expert metallurgists, none of whom was a member of the core revalidation team.

> **Example – Consolidated Facility Siting Study**
>
> Some facilities have a consolidated, quantitative facility siting study that applies to the entire location. While these facilities would reference the study in the individual unit PHA revalidations, the consolidated study may or may not be *Updated* along with each revalidation.
>
> It is prudent for the facility to collect changes identified in the individual unit revalidations and periodically *Update* the consolidated facility siting study to ensure that it is accurate and that new buildings, release points, and siting concerns have been considered.

Redoing the Complementary Analyses. Complementary analyses should be evaluated just like core studies in deciding whether they must be *Redone*. However, few organizations require that complementary analyses be *Redone* simply because of their age. The *Redo* approach is usually selected when there are significant changes in many conditions affecting the risk judgments in the complementary analyses, such as factors impacting human error rates (e.g., procedures, lighting, labeling) or facility siting (e.g., hazardous inventories, control rooms, emergency response equipment). When specific circumstances warrant, they typically apply to only one or two of the complementary analyses, not all of them. A *Redo* of a complementary analysis typically involves performing the analysis as if it were a new or initial analysis following the specific rules and guidelines for that analysis.

Two of the most common complementary analyses address specific concerns associated with facility siting and human factors. Example facility siting checklists are provided in Appendix E, human factors checklists are provided in Appendix F, and an example external events checklist is provided in Appendix G. If blank checklists such as these are to be *Redone*, the revalidation team reads each question, typically gives a "Yes," "No," or "Not Applicable" answer, and then provides a summary of any pertinent discussions associated with each question or topic. New recommendations are made when the team concludes that improvements can be made or when deficiencies are identified. Some facilities limit these checklist recommendations to items that are a specific process safety issue or loss of containment concern, while other organizations allow any recommendation that would lead to process or process safety improvements.

Updating the Complementary Analyses. The *Update* approach is usually selected when there are few changes in conditions affecting the risk judgments in the complementary analyses. Revalidation of the core analysis often triggers the need for specific *Updates* of the complementary analyses. For example, the decision to receive a raw material in railcars instead of trucks might require an *Update* to the facility siting study. An incident involving hydrogen embrittlement of a specific alloy might require an *Update* of the damage mechanism review. Many of the circumstances warranting *Update* of the complementary analyses become apparent during the revalidation of the core analysis. Thus, the revalidation team should explicitly recommend *Updates* to the complementary analyses as it deems appropriate. Then, the appropriate technical experts will *Update* the complementary analyses as necessary.

Facility siting and human factors checklists can also be *Updated* if they are of good quality and are complete. If the existing facility siting checklist will be *Updated*, the same checklist used in the prior PHA is reviewed and modified as necessary by the current revalidation team. Responses are updated based on new operating experience, changes, and incidents, in a similar manner to *Updating* the core analysis.

Typically, for a **facility siting** checklist *Update*, these discussions include asking questions such as:

- Were there changes to the process or process structures that have:
    - o Adversely affected unit layout or spacing between process components?
    - o Adversely affected the location of large inventories of hazardous materials (e.g., changes in normal inventory levels; receipt of material in larger containers; temporary storage of material in totes, drums, trucks, or railcars)?
    - o Introduced any new hazardous materials?
    - o Adversely affected Motor Control Centers?
    - o Adversely affected the location or construction of the control room or any other temporarily or permanently occupied buildings (e.g., engineering, maintenance, laboratory, administration)?
    - o Resulted in any new hazardous events, or significant changes to previously identified hazardous events, with the potential for impacting occupied buildings? (within the PHA boundary, within the facility, or outside the fence line)
    - o Added likely sources of ignition?
    - o Introduced new electrical area classification issues?
    - o Impaired the coverage of the firewater system or its activation?
    - o Adversely affected the adequacy of drains, spill basins, dikes, or sewers?
    - o Impaired access to, or warranted, installation of additional, emergency stations (e.g., showers, respirators, personal protective equipment [PPE])?
    - o Created new facility siting concerns that should be updated into the existing facility siting study for the entire facility?
    - o Created the need for additional passive fire protection or impaired existing protection (e.g., for support structures)?
- Have there been any changes outside the PHA boundary that could adversely affect facility siting of this process? (e.g., new construction of other operating units, new construction near the fence line, new release points, or discharge locations)
- Has operating experience since the prior PHA identified any new hazardous events with the potential for impacting occupied buildings?

Typically, for a **human factors** checklist *Update*, these discussions include asking questions such as:

- Are housekeeping and the general work environment the same or have they been improved?
- Have there been any changes in the accessibility and availability of controls and equipment?
- Is the equipment labeling still accurate and well maintained?
- Have there been any changes in the feedback/displays of controls?
- Have there been any changes in personnel workload/stress (e.g., due to changes in staffing levels, experience or competency, assignment of new duties)?
- Have the procedures been updated to reflect changes, and have they been field validated?
- Have workers been trained on changes, and is refresher training up to date?

Using both *Redo* and *Update* approaches is conceivable for complementary analyses that span the PHAs for multiple units. For example, a comprehensive facility siting study might have to be *Redone* for one particular unit and *Updated* for all the others. In that case, the experts may solicit more specific scenario information from the core revalidation team, but the conduct of the revalidation is essentially the same as an *Update*.

7.1.3   *Revalidation of Supplemental Risk Assessments*

The core analysis of any PHA includes at least a qualitative risk assessment. The mere fact that the PHA team made any recommendation implies that they concluded the "as found" risk exceeded the organization's tolerance or that there were other opportunities for improvement. However, past judgments have proven to be highly subjective, depending on the individual team members' personal knowledge, experience, and interests. Thus, many organizations now require that their PHAs include quantitative risk analyses that are more consistent within a particular study, across all their PHAs, and across the entire organization. Often, only an order-of-magnitude risk analysis is required to align the PHA team's risk judgments with the organization's risk tolerance as expressed in its order-of-magnitude risk matrix. These simplified quantitative risk analyses, such as LOPA, follow a conservative set of simple rules that are standardized across the organization.

The revalidation approach for supplemental risk assessment generally tracks the approach chosen for the core analysis, but it is a one-way relationship. If the core analysis must be *Redone*, then the supplemental risk assessment must also be *Redone*. However, it is possible the supplemental risk assessment will need to be *Redone*, even if the core study was only *Updated*. Supplemental risk assessments may be performed as part of the core analysis revalidation and are coordinated with it. Some organizations select a more specialized team to perform the supplemental risk assessments, and their work is usually scheduled shortly after the

> **Example – An Updated PHA Scenario Requires Updating a LOPA (or any other supplemental risk assessment)**
>
> The core analysis might, for example, be *Updated* to include a new alarm that had been installed. If the supplemental risk assessment using the LOPA technique was applied, the new safeguard (1) might not qualify as "independent," and the risk analysis would not change or (2) it might meet the company independent protection layer (IPL) requirements and be credited in the LOPA. Similarly, if a Human Reliability Analysis had been performed instead of LOPA, it should be *Updated* to include the new alarm.

revalidation of the core analysis. Some key members of the revalidation team are usually part of the specialized team to expedite the risk discussions and help ensure consistency with the core analysis.

Redoing the Supplemental Risk Assessments. The *Redo* approach is usually selected when there are significant changes in many conditions affecting the risk judgments in the supplemental studies. *Redoing* the core study is the predominant reason to *Redo* the supplemental risk assessment, and if so, efforts should be coordinated, just like an initial PHA. However, if only the supplemental risk assessment is being *Redone*, perhaps due to changes in the company LOPA procedure, then the execution timing could be different. In that case, it may be more efficient for the risk specialists to *Redo* the supplemental risk assessment of the entire study first. Then, the core study revalidation team can proceed to *Update* the core study and *Update* the supplemental risk assessment as discussed in the next paragraph.

Updating the Supplemental Risk Assessments. The *Update* approach is usually selected when there are few changes in conditions affecting the risk judgments in the supplemental studies. Supplemental risk assessments should be *Updated* in lockstep with *Updates* to the core analysis. With each *Update* to the core analysis, the revalidation team should ask whether the change will affect the supplemental risk assessment, and if so, note the interconnection.

Conversely, it is equally important to document which core analysis items the supplemental risk assessments stem from.

Using both *Redo* and *Update* approaches is common for supplemental risk assessments, and they typically follow the approaches of the core study revalidation. Risk assessments of the *Update* portions are *Updated*, and risk assessments of the *Redo* portions are *Redone*. Alternatively, a conservative approach would be to *Redo* all of the supplemental risk assessments, using the previous assessments as a guide where practical.

7.2 FACILITATING EFFECTIVE REVALIDATION MEETINGS

The skills that trained PHA facilitators acquire on the job are as applicable to the efficient conduct of a PHA revalidation as they are to the conduct of an initial PHA. PHA team facilitation is discussed in detail in the CCPS book *Guidelines for Hazard Evaluation Procedures* [2]. The leader, however, should consider some unique aspects of a revalidation session. Most fundamentally, the revalidation team has the previous PHA documentation to use in a manner consistent with the *Update* or *Redo* approach. At the very least, "holes" in the previous PHA documentation show where extra effort needs to be applied in the current revalidation. In addition, the revalidation team may have the benefit of several years of actual operational and maintenance experience in lieu of the original design team's optimism. Attention to items discussed in the following sections will help the facilitator keep the sessions as productive as possible.

> **Inexperienced Team?**
>
> Team makeup is often a management decision outside of the revalidation team scope. The revalidation leader, however, should make responsible managers aware of a particularly inexperienced team.
>
> Such a team could lead to a deficient revalidation or an artificially high number of recommendations involving the conduct of additional studies or evaluations on issues that should, more properly, be resolved by the revalidation team during the meetings.
>
> An experienced team is critical to the success of any revalidation.

7.2.1 Team Composition

Core members of the revalidation team, meeting all the regulatory and company policy requirements, should be selected prior to the study sessions, as was discussed in detail in Section 6.1.2. Beyond that, subject matter experts (SMEs) should be selected to assist the core team in evaluating specific topics, as needed. It may be possible to organize the review sessions and selectively invite specific SMEs to join the core team for discussion of topics in their particular area of expertise. For example, if the session topic is a review of maintenance work orders to determine the significance of any changes identified, the quorum for this session might not include the laboratory supervisor or the control engineer responsible for the distributed control system. If the focus of the session is to evaluate changes in the operating procedures, involvement of the maintenance supervisor or inspection supervisor might not be necessary.

SMEs outside the core team are typically only required to participate in sessions related to their expertise. As long as regulatory and company policy requirements are met, such selective involvement can ensure better use of resources and increased personnel interest and motivation.

7.2.2   Meeting Kickoff

The PHA revalidation opening meeting is the best time to introduce the team members and discuss the significance of the PHA. In addition to the PHA revalidation team members, it is often helpful to include the manager responsible for the PHA being revalidated and various support team members who will be available for reference during the sessions, but not always present during the meetings. These special attendees, who usually participate as needed, can include facility management, mechanical and electrical engineers, process controls experts, laboratory personnel, industrial hygiene experts, contractors, vendors, and other SMEs. Any team members who will be participating remotely should also be included in this meeting. Using a checklist similar to the example in Table 7-1 can help ensure an effective kickoff meeting.

After the customary introductions, the study leader (often called the "facilitator") should clearly state the scope, purpose, and schedule of the revalidation. The leader should also state the ground rules for meeting discussions (sometimes called the code of conduct), to establish expectations for an orderly, respectful meeting. As in any PHA meeting, the goal is to engage all the team members in productive discussions (e.g., by not allowing a few people to dominate, by asking direct questions to individuals, by using familiar terminology). This is particularly important for revalidation teams that include members of the prior PHA team who may feel that they are being personally attacked or that their judgment is being questioned. Team members may be encouraged to say, "What I heard is..." instead of "You said..." All this information should be summarized in an easily visible location for the team members (e.g., on a flip chart or marker board for in-person attendees or electronically), so all the team members have a common understanding.

Table 7-1 Sample Kickoff Meeting Checklist

Item Completed	Topic
	Introductions/Schedule
	Revalidation Scope
	Team Training
	Meeting Rules or Revalidation Team Charter
	Revalidation of PHA Core Methodology
	Consequences of Interest
	Previous Incidents
	Complementary Analyses (checklists to be used)
	Risk Tolerance/Risk Matrix
	Supplemental Risk Assessment Tools (e.g., LOPA)
	Recommendations from the Prior PHA
	Unit Tour
	Other Topics for Review and Questions

The study leader should first explain the physical scope of the review. That description may be as simple as, "everything within the unit battery limits," or as specific as enumerated pieces of equipment. If there is a difference between the minimum scope required by regulation and by company requirements, the leader may need to identify where the team will be going beyond regulatory minimums. This distinction will help team members understand why they may observe other differences, such as the MOC records, the selection of scenarios for LOPA, or the priority assigned to recommendations. A highlighted process flow diagram is a useful visual aid for this discussion, and it is something tangible for later reference.

The leader should also explain the operational and analytical scope of the revalidation. In addition to normal operation, the leader should explain how to address the hazards during other operating modes, such as startup or maintenance. Depending on the core analysis method, there may be specific checklists, questions, deviations, or nodes applicable to other operating modes. Regardless, the leader's primary purpose is to engage the team in identifying process hazards in any mode of operation within the scope of the revalidation.

If some team members have not previously participated in a PHA, the leader will need to explain the purpose and methodology of the PHA. (See Chapter 1.) Some personnel may not be accustomed to the critical analysis process used in a PHA or may be unfamiliar with the cause/consequence/safeguards logic of the PHA process. These concepts should be reviewed with the team members, or they should be appropriately trained, before the revalidation begins. Even if the team members have some PHA experience, a skilled facilitator will typically lead the team through a period of "on-the-job training" after the kickoff meeting to reinforce their understanding of the core analysis method, to reinforce the ground rules for team discussions, and to reconcile the diverse experiences team members may have had in other studies.

The study leader should then explain the selected revalidation approach (ranging from a complete *Redo* to a focused *Update*) as described in Chapter 5. Team members will be more committed to the revalidation activity if they understand the reasons a specific approach was selected, particularly if they perceive the selected approach will involve additional or significant work for them. Since a PHA was done previously, some revalidation team members may be prone to believe that their time is being misused, and that belief gets stronger the closer the revalidation gets to a complete *Redo*.

Therefore, it is crucial that the facilitator immediately discuss with the team members that this is a good opportunity to reduce process safety risks that could catastrophically affect them, their colleagues, their community, and the environment. Further, the team leader should review the unit's history of process safety incidents, with emphasis on any incidents occurring since the last PHA. This review proves there are real opportunities to improve performance, and it helps rebut any later claims during the meeting discussions that "That could never happen here!" or "We would never do that!" (or something very similar).

During the kickoff meeting, the study leader should also explain the risk tolerance criteria and consequences of interest to be used. The team leader should discuss any differences with criteria used by the prior PHA team, again emphasizing the need for a thorough but efficient review.

Lastly, and particularly if the *Update* approach will be used, the study leader should initiate discussion of the change history of the unit. The goal is not to review specific hazards in detail, but to give the team members a sense of the revalidation's scope and complexity. It also gives the leader an opportunity to solicit the team members for information about any additional changes or problem areas that were not mentioned.

Once the study leader resolves any issues that may arise (e.g., information access and exchange with remote team participants), the revalidation can commence. However, before the actual meetings begin, some companies allocate time and resources for teams to tour the unit together. Such a tour would be of greatest value to team members who do not routinely work in the unit and who did not have an opportunity to visit the unit before the kickoff meeting. If not explicitly budgeted and planned, such a tour takes time away from the team's scheduled meetings. Practical considerations of access permits, escorts, PPE, weather, noise, disruption of other unit activities, etc. must also be considered. Thus, the team must make a judgment as to whether the cost of a group tour is worth the benefit for the team as a whole, whether other arrangements should be made for particular individuals, or whether tours will be conducted as needed during the revalidation sessions. Sometimes a virtual tour is possible with three-dimensional photography or animation. Regardless, a tour will help ensure a common familiarity with the process and a common frame of reference with respect to facility layout and equipment location. It will also help identify and correct any misconceptions that team members might have as to the physical scope of the review. Many revalidation teams have found the benefits of a field unit tour are well worth the time invested.

7.2.3   Meeting Productivity

Meeting time is one of the team's most precious resources, and one of the facilitator's main goals is to optimize the use of that time. For example, there is a significant amount of preparatory work involved in locating, collecting, and preparing information for use in the revalidation sessions. As discussed in Chapter 6, it is incumbent that the study leader, or designee(s), do the needed preparatory work ahead of the meeting so that this information can be accessed and used as efficiently as possible. Piping and instrumentation diagram (P&ID) revisions can be cross-referenced to PHA nodes, and to MOC and pre-startup safety review (PSSR) documentation, before the sessions begin. Sessions will be more productive, and less onerous for participants, if documents are organized so they can be found quickly and efficiently when they are needed.

A unique issue in a revalidation meeting, particularly one using the *Update* approach, is the use of the previous PHA. The monotony of the leader simply reading it to the team or flashing it up on a screen can easily lead to the team becoming disengaged and result in an incomplete or otherwise inaccurate analysis. On the other hand, forcing the team to regenerate accurate information already in the existing documentation is of little value. The facilitator's challenge is to keep the team mentally engaged while maximizing the team's efficiency. Often this can be accomplished with simple changes in language. Rather than display text from the previous PHA and ask the team, "Good?" or "Does that look right?" the leader might ask questions such as:

- Can you think of any other causes?
- How would an operator diagnose that?
- Have you had any trouble with those alarms?
- How might those safeguards fail?
- Did the resolution of this recommendation solve the problem?

When the revalidation is an *Update*, the facilitator can usually engage the team's interest by reviewing MOCs/PSSRs as one of the first activities. If possible, start with an MOC that involves an entire node or requires the creation of a new node because (1) it is something new and interesting and (2) it allows the leader to refresh the team on terminology (e.g., causes, consequences, safeguards) and techniques (e.g., HAZOP, What-If/Checklist, FMEA) that will be used throughout the rest of the study. The review of MOCs and PSSRs can then be naturally extended to include discussion of any incidents or recommendations in the prior PHA that did *not* result in an MOC. If the revalidation is a *Redo*, MOCs and incidents can be reviewed near the end of the study as a check on the thoroughness of the assessment.

The following examples describe situations that the facilitator should avoid, actions they can take in response, and benefits to educating the team during the revalidation meetings. By being aware of these types of situations, a facilitator can help keep the revalidation meetings on track and help ensure the quality and completeness of the analysis.

Example 1 – Understating Potential Consequences

Team members (prior or current) often mistakenly take credit for safeguards when describing potential consequences, because it aligns with their work experience and expectations. For example, if the prior PHA describes the consequence of overpressure significantly above maximum allowable working pressure (MAWP) as "discharge to scrubber," they took credit for the expected action of the pressure relief valve; the potential consequence is almost certainly "vessel leak" or "vessel rupture." The leader should seize such opportunities to highlight the error to the revalidation team so they can be vigilant for similar errors in the prior PHA and in their own discussions.

Example 2 – Incorrectly Changing Potential Consequences

Team members sometimes mistakenly claim that mitigation safeguards reduce the severity of consequences rather than their likelihood. This is a difficult concept for some team members because it seems to defy logic. By definition, a mitigation safeguard, such as a fire sprinkler system, reduces the expected consequences. Yes, but a building with an automatic sprinkler system can still burn to the ground if the sprinklers do not work. Therefore, the correct risk analysis is that the sprinklers reduce the likelihood of burning the building down. If the leader can spot an example of how the prior PHA team mistakenly judged risk tolerable by moving down the severity axis, rather than down the frequency axis of their risk matrix, it can be used to explain the correct use of a risk matrix and facilitate future discussions during the revalidation.

Example 3 – Incorrectly Using Safeguard Failures as Causes

Team members sometimes mistakenly claim that failure of a safeguard is the cause of the hazard it is designed to protect against. For example, the prior PHA may claim "relief valve fails to open" as the cause of overpressure. This is not true. The relief valve could be welded closed, and there would never be an overpressure until something else caused it. A pressure relief valve safeguard could fail such that it does not protect against overpressure, but it did not cause it. The leader should highlight such errors at every opportunity to coach the team and enlist their assistance in finding and correcting similar mistakes.

Example 4 – Incorrectly Judging Cause Frequency

Team members sometimes misapply site experience to justify lower cause frequencies. For example, consider a pressure control valve that could malfunction and cause overpressure. A team member might say that there has never been an overpressure and, therefore, the frequency of pressure control valve malfunctions must be lower than the prior team assumed. The team leader should challenge that assertion by asking how many times the relief valve has opened. It is likely that the team member does not remember any overpressure incidents because the relief valve operated correctly and prevented any damaging overpressure. The opposite may also occur. If the prior team was using typical LOPA rules for initiating event frequencies, they may have claimed the valve failure frequency was once in ten years. However, if the revalidation team knows the valve has failed several times, the facilitator should document that a higher frequency based on actual field experience in this process is being used rather than values from the default LOPA frequency table. In either case, the facilitator should use the opportunity to explain basic statistics to the team.

Typically, a small number of negative experiences are needed to justify increased cause frequency, but a lot of positive experience is needed to justify a decreased frequency.

Example 5 – Inconsistently Documenting Loss Scenarios

Maintaining consistency is another constant challenge, particularly when the *Update* approach is used. For example, the revalidation team may want to list the new check valve, installed as recommended by the prior PHA team, as a safeguard against backflow. However, check valves are not listed elsewhere in the prior PHA as safeguards. Alternatively, the revalidation team may believe the consequences of a particular chemical release from a new vessel are very severe. However, similar chemical releases in the prior PHA were categorized as lower severity. In both cases, the revalidation team must decide whether to maintain the precedents in the prior PHA and continue the limited *Update* as planned, or to substantially increase the scope of their efforts to review every instance and appropriately *Update* the PHA everywhere their judgment conflicts with those documented in the prior PHA.

> **Identifying Documentation Inconsistency**
>
> Some commercially available PHA software packages make it easy to generate custom reports that gather scenarios with the same severity ranking or same risk ranking. The team leader can use this tool to review the worksheet entries and identify descriptive entries that seem inconsistent with others in that group of scenarios. For example, a scenario may have a severity ranking implying fatality potential, but the consequence description only speaks to equipment damage or a minor operational issue.

Example 6 – Designing Solutions

Similar to an initial PHA, the purpose of a revalidation is to judge current process risks, NOT to provide specific design solutions if elevated risks are found. Team members often have excellent ideas on how risks might be effectively reduced, and the facilitator wants to elicit those ideas. However, the discussion can quickly devolve into an argument about which solution is "best" and waste precious team meeting time. The facilitator should firmly intercede, summarize the ideas as a functional recommendation, and move the team forward. For example, the facilitator might say, "The team seems to have concluded that the risk of overfilling the tank exceeds our tolerance. I have noted our recommendation to reduce the risk of overfilling Tank 101 by means such as adding a high-high level alarm, flow totalizers, weigh cells, or by enlarging the containment dike. If there are no other ideas, let us move on."

Example 7 – Excessive Conservatism

Risk analysts tend to be conservative. Thus, revalidation teams have a tendency to recommend additional safeguards, even when the risk is categorized as tolerable. Given finite resources, the resolution of any recommendations to reduce tolerable risks further will consume resources better spent on resolving recommendations to reduce elevated risks. Furthermore, a recommendation to reduce one particular risk may be more than offset by increases in other risk(s). For example, adding a redundant pressure relief valve may marginally decrease the risk of vessel rupture due to overpressure. However, the additional relief valve might fail open under normal operating conditions, significantly increasing the risk of a loss of containment when no overpressure exists. Thus, the study leader should challenge recommendations to add more safeguards when risk is already tolerable.

Example 8 – Excessive Optimism

Team members tend to be optimistic in their risk judgments, particularly when human performance gaps are involved. In their experience, workers make very few serious errors, and when they do, they detect and correct the error themselves, or the engineered controls act to minimize any harm. If a particular revalidation team is knowledgeable of LOPA, it may be difficult for them to accept that a particular human error could happen more than once in ten years. A facilitator should watch for phrases like "no one would ever close the wrong valve around here." A review of past incidents or a discussion of these types of events with operators might reveal these, or similar, events do indeed happen occasionally. The team might then consider a more detailed supplemental risk assessment using Human Reliability Analysis techniques where applicable.

This general tendency towards optimism also extends to equipment performance. Many LOPA teams work hard to get a loss scenario to show a tolerable risk (sometimes by taking credit for too many basic process control system [BPCS] safeguards such as IPLs, double counting IPLs, or ignoring common cause failures between IPLs). If such issues are noted, the leader should challenge the team to follow their revalidation charter and recommend risk reduction measures whenever they determine that elevated risks exist, as estimated using company-approved methods.

Example 9 – Defending Prior Work

Study leaders should try to engage everyone in the team deliberations. This is particularly true in a revalidation because some of the team members may have a personal investment in the prior work (PHA, MOCs, incident investigations, etc.) that is being reviewed. For example, the engineer on the revalidation team may have been the person responsible for an MOC being examined. They naturally know more about the details of the MOC and its conclusions than anyone else does, so the team will let them speak first. However, the leader cannot let that explanation dominate or eliminate any further discussion. If no one else speaks up, the leader should direct specific questions to other team members, such as asking the operator whether the change resulted in any alterations to the startup procedure or asking the maintenance expert whether the change affected how the equipment is isolated and drained.

Revalidation leaders should also intervene if they sense that team members know that the recommendations needed to reduce risk will not be well received or that they do not want to face the work that the recommendations will generate. Challenging prior risk judgments may provoke a confrontation with other team members who made those judgments and damage team cohesion. Team leaders should immediately redirect any questions about individual competence or motives in the past to the current factual issue(s), considering risk in light of what is known today. To move the revalidation forward, the leader may suggest that a contentious issue be resolved using a company-approved method such as LOPA. If LOPA confirms the risk is elevated with respect to the organization's current risk tolerance, the leader can then solicit team ideas for risk reduction.

Example 10 – Assuming Similar Process Equipment Has Identical Risks

Where multiple pieces of equipment or multiple trains are nominally the same, a common PHA practice is to analyze one component or train as typical of all. That may have been true in the original design. For example, all the reactors may have been equipped with identical high pressure/temperature alarms. However, the alarm setting may have subsequently been changed for operational reasons, and the changes were not all the same. When the facilitator encounters nodes that were previously analyzed as "typical" of similar equipment, the facilitator should closely question the revalidation team to ensure that the equipment has no differences that would significantly affect the risk judgments.

7.3   REVALIDATION MEETING CONCLUSION

Some revalidation team leaders find it effective; after all other items are discussed, to guide the revalidation team through a period of general discussion to ensure that no significant topics were overlooked. These general discussions help identify changes in global risks that may have been missed during previous discussions, such as loss of instrument air to the entire process unit or node, or the impact of external events (e.g., floods, wildfires, or blizzards).

Additionally, the general discussion can be used to ensure that the PHA revalidation complies with applicable company or regulatory requirements by specifically reviewing and documenting discussions on required considerations (often through use of a PHA completeness checklist like that in Appendix B). Such considerations could include human factors, facility siting, and previous incidents. Note: These topics should have been addressed earlier in the PHA, but some leaders like to revisit them at the conclusion of the study to ensure their comprehensive treatment.

Finally, the general discussion may be used to document discussion of issues that are common to multiple study sections, such as corrosion monitoring programs or relief header/flare sizing. Some areas of discussion may be enhanced if specialist personnel with particular expertise participate.

Companies may wish to develop a checklist, specific to their particular needs, for use in guiding these discussions. Table 7-2 is an example checklist that can be used alone or as part of a more comprehensive procedure for concluding the PHA revalidation sessions. If review of any of the items initiates thoughtful discussion (or identifies missed topics in the meeting), the team should return to pertinent locations in the study and address these items to completion. These discussions may also identify risks that were not within the scope of the revalidation, such as adversary threats. The team may recommend a separate study or studies to address issues such as cybersecurity, physical security, espionage, or product tampering.

Table 7-2 Sample General Topic Checklist

Item Discussed	Topic

	Safety and fire
	    • Fire protection systems
	    • Area electrical classification
	    • Industrial hygiene

	Emergency response
	    • Community emergency response plan
	    • Plant/facility emergency response plan

	Global loss of utilities
	Facility siting
	Human factors

	Mechanical integrity
	    • Testing and inspection
	    • Training
	    • Damage mechanism reviews

	Maintenance

	Recommendations from other PSM activities
	    • Previous PHAs
	    • Incident investigations
	    • Emergency response drills
	    • Audits

	Environmental requirements
	External events
	Procedure review (e.g., PHA or LOPA procedure)
	Other company requirements
	Other local requirements

To conclude the meetings, some study leaders review the complete list of recommendations with the revalidation team. Risk judgments may have changed, given the totality of the discussions, and recommendations may need to be added, broadened, narrowed, or deleted entirely. This finalized list is the most valuable product of the revalidation team's efforts that will be documented in the revalidated PHA.

7.4   PRINCIPLES FOR SUCCESSFUL REVALIDATION MEETINGS

Actual experience in conducting revalidations, by PHA practitioners across a number of companies, has highlighted some keys for success, as well as some things that can impede success. While none of the items listed are absolute rules, they do provide valuable guidance.

Successful Practices:

- Having a formal code of conduct or team charter that fosters participation by all team members
- Ensuring the team includes members with all the required skills and diverse experience
- Including process experts who will be available as needed for the revalidation sessions and include them in the opening meeting
- Informing the team members of the revalidation strategy for each aspect of the PHA (core, complementary, and supplemental analyses) and explaining the rationale for each
- Refreshing the team members' understanding of the organization's risk tolerance criteria
- Ensuring that some team members were not involved in performing the prior PHA
- Preparing appropriate worksheets to capture revalidation team discussions
- Providing team members with an overview of process operations, an overview of the physical layout, and a refresher in the analysis techniques, as necessary
- Providing team members with an overview of changes since the prior PHA
- Suspending a team session when an adequate team is not present and rescheduling a later time when required participants (or substitutes with similar qualifications) will be available
- Providing team members with an overview of recent, relevant, process safety incidents
- Performing a detailed review when *Updating* the prior PHA. A detailed review affords the PHA revalidation team the time and resources to help identify new concerns/hazards that may have been overlooked in the prior PHA

- Reviewing specific changes (especially changes made due to recommendation or incident) since the prior PHA when performing a *Redo*. Important insights may be missed if changes resulting from incidents are not specifically evaluated for their effectiveness in preventing recurrence
- Fully describing the scenarios, including the intermediate events that led to the consequences, so that future revalidation teams can understand the current PHA team assessment
- Using the experience of the PHA team to revalidate supplemental risk assessments (e.g., LOPA/QRA), as well. The team can provide insight as to the functionality and reliability of critical safeguards used in these assessments and *Update* them accordingly
- Being vigilant for unresolved recommendations from the prior PHA
- Using checklists for special or unique requirements that apply to processes to remind PHA revalidation team members of key points to verify

Obstacles to Success:

- Using an inadequate team and/or facilitator expertise
- Failing to obtain stakeholder agreement on the purpose, scope, methodology, and schedule prior to the first team meeting
- Repeatedly using the same study leader and/or team to *Update* the PHA of a particular unit
- Relying on team members with too much personal investment in the prior PHA (e.g., they do not think anything should be changed or that any recommendations should be made), despite evidence of inaccuracy or inconsistency
- Failing to manage conflicting priorities impacting the team meeting schedule and the participation of required personnel
- Using inadequate PSI (e.g., incomplete P&IDs or operating procedures)
- Failing to revalidate complementary and supplemental analyses, in addition to the core analysis
- Failing to consider past incidents and operating experience, including those incidents and changes that were not controlled
- Failing to evaluate how adequately the recommendations of the prior PHA were resolved
- Failing to maintain consistency with the prior PHA when *Updating* causes, consequences, safeguards, or risk rankings

- Trying to force an *Update* of the PHA when a *Redo* is necessary. *Updating* a poor PHA can result in a poor revalidation product, without saving time or resources
  - o The best course of action after making this revelation during the PHA sessions is usually to reconvene the team after the study leader has prepared for a *Redo*
- Failing to consider the accuracy of previous PHA entries during an *Update*. This can allow errors to perpetuate in a PHA
- Underestimating the meeting time requirements prior to the sessions, and therefore not having enough calendar or meeting time scheduled for the revalidation
- Failing to perform the core methodology appropriately. For example:
  - o Not considering causes if they originated in equipment or processes beyond the physical boundaries of the PHA being evaluated or taking credit for safeguards when evaluating consequences
  - o Claiming invalid or ineffective safeguards or rejecting plausible scenarios as non-credible (i.e., "that has never happened here")
  - o Using the same safeguards repeatedly without any real thought as to their effectiveness or specific application to a scenario (e.g., overuse of "Operator Training and Knowledge")
  - o Failing to detect common cause failure mode issues between process control and safety/emergency shutdown systems
  - o Inconsistently ranking the risk or consequence severity of similar scenarios
  - o Being overly optimistic in estimating consequences or likelihood, resulting in an underestimation of risk
- Failing to address facility siting, human factors, external events, or other required topics (typically as a complementary analysis)
- Making recommendations that are either too vague or impractical to implement

8 DOCUMENTING AND FOLLOWING UP ON A PHA REVALIDATION

A clear, concise, thorough PHA revalidation report is essential to the retention and communication of the PHA results, as well as being needed to demonstrate compliance with internal and external PHA requirements. Because PHA is the PSM element through which an organization identifies process hazards and appropriate risk controls, a revalidated PHA report provides an ongoing basis for many other PSM program elements. For example, it identifies the engineered controls that must be included in the mechanical integrity (MI) program for inspection, test, and preventive maintenance (ITPM) to maintain the required probability of failure on demand; it identifies the loss scenarios that must be addressed in the emergency response plan; and it identifies the consequences of deviations that must be addressed in the standard operating procedures. It is also the basis for future management of change (MOC), incident investigation, and revalidation activities, and it is subject to periodic audits that verify compliance with regulatory and/or internal company PSM requirements.

Experience shows that inadequate documentation of the prior PHA is one of the most frustrating issues for the future revalidation team because (1) it causes unnecessary work that consumes additional time and distracts the team from its goal of evaluating hazards and risks and (2) it may require them to *Redo* a PHA that could otherwise have been *Updated*. It is likely that the revalidation team will gain, during the course of the PHA revalidation, a greater appreciation of the importance of complete, accurate documentation and document their own work accordingly.

Much of the documentation for a PHA revalidation parallels the typical documentation for an initial PHA (e.g., including a current set of drawings showing the node definitions). However, some approaches to revalidation, as suggested in this book, use a number of screening forms and checklists. Those choosing to use forms and checklists (such as those in Appendices A and B) should consider including them in the revalidation documentation so that the basis for key decisions (e.g., the basis for the particular revalidation approach chosen) can be clearly communicated to the next revalidation team.

The PHA revalidation report format and content may be prescribed by facility or company requirements. This chapter provides suggested documentation practices for the PHA revalidation and associated records that could be used, if specific local requirements are not provided.

8.1   DOCUMENTATION APPROACHES

The spectrum of approaches to revalidating a PHA naturally results in a spectrum of documentation styles. For convenience and consistency, the terms *Redo* and *Update* are used herein to compare and contrast the documentation that typically results from those approaches. But, just like the revalidation activities themselves, the documentation styles are not mutually exclusive, and practical applications usually result in a combination of both being used for different portions of the PHA. It is possible that a team using the *Update* approach might copy all relevant information from prior PHA(s) into the current report, so the reader would not need to refer back to the older documents. In that case, the *Update* analysis would be documented in the *Redo* style.

With *Redo* documentation, a new and detailed PHA report, similar in format and content to the initial PHA report, is prepared. This single document:

> **Terminology Note**
>
> In the first edition of this book, the terms *"basic"* and *"evergreen"* were used to describe revalidation documentation styles. Those terms were eliminated in this edition and replaced with *Update* and *Redo*, corresponding to the revalidation approaches. Some organizations may continue to use the older terms because they are embedded in their internal guidance. The term "evergreen" mentioned in Section 1.6 refers to an ongoing revision of the current PHA analysis worksheets, NOT a revalidation.

- Identifies the process unit examined and the reason(s) various portions of it were included in the PHA (e.g., Nodes 1-26 are required by national regulations; Nodes 27, 32, and 41-48 are required by local regulations; and all other nodes are required by company policy)
- Lists meeting participants and who, specifically, filled roles required by regulation or policy. Some organizations include a log of the activities that occurred on each meeting day

- Lists documents examined (e.g., prior PHA report, piping and instrumentation diagrams [P&IDs], MOCs and incidents since the prior PHA)
- Lists key dates (e.g., meeting dates, freeze date for data inputs into the revalidation, report issue and approval dates)
- Describes the revalidation approach and the rationale for its selection
- Documents the core analysis and its recommendations
- Documents or references any complementary analyses (e.g., a human factors checklist, human reliability study, or facility siting checklist) with their recommendations
- Documents any supplemental risk assessments (e.g., LOPA or bow tie) with their recommendations

Appendices to the *Redo* report could contain the core analysis worksheets (e.g., HAZOP worksheets), other supporting documentation such as the Change Summary Worksheet in Appendix C, any complementary analyses (e.g., checklists, studies) used by the revalidation team, and/or any supplemental risk assessment worksheets.

Redo documentation should result in a report that:

- Integrates and presents the requisite information in a format that will be familiar to users
- Simplifies revalidation in the future
- Demonstrates that all analysis requirements were met and that any issues with the prior PHA were addressed
- Supports other process safety activities (e.g., MI, MOC, training)
- Accurately describes the current process configuration, hazards, and applicable risk controls
- Provides enough information so the justification, context, and nature of the recommendations can be easily understood by those personnel, including possible contractors, who will be responsible for follow-up resolution and implementation activities

The *Redo* documentation then becomes the complete and current baseline for all ongoing PSM activities that rely on the unit PHA. Where practical, the *Redo* documentation should include relevant, supporting information for the study, including copies of any supplemental analyses (e.g., consequence analysis computations/reports) done as part of prior PHA cycles, to allow access to the important information without having to refer back to previous reports.

However, it is not necessary for the PHA to include copies of documentation controlled by other management systems. For example, if the procedures reviewed were version controlled, the revalidation report would only need to provide the document title and revision number. The goal is to provide sufficient traceability so that the documents used by the revalidation team could be retrieved, if necessary.

The *Update* documentation style preserves the prior PHA as one of the input documents. In this documentation style, the previous *Redo* PHA report is required along with the *Update* report to explain the scope and methods used in the analysis. The *Update* report, similar to the PHA *Update* approach itself, explains what was reviewed and changed in the *Update*.

The revalidated report may look significantly different if the team performed an *Update* by "detailed review," as described in Chapter 5. In that case, the revalidation report may include new or substantially revised worksheets and may have significantly revised the body of the report itself. If the revalidation team performed a "change and incident" review, then the worksheet documentation in the PHA might look essentially the same.

An *Update* Caution

The simplified *Update* documentation approach may make it more difficult for others to use the report(s). This is because it is not always integrated and streamlined into a single report like the *Redo* format, and the difficulty of use increases exponentially with each additional revalidation cycle that is *Updated*. This is particularly true if the "simple" *Update* approach is used repeatedly. Therefore, similar to the decision to *Update* or *Redo* the PHA itself, it is a good practice to periodically *Redo* the PHA documentation (likely on the same PHA *Redo* schedule).

Additional individual documents (primarily any MOC forms, incident investigation reports, and/or PHA recommendation resolutions completed since the prior PHA) are typically added, along with any worksheets or checklists completed during the revalidation. A cover report then summarizes the revalidation process and methods (similar to the bulleted items in the *Redo* report discussed previously in this section), the conclusions from the core and complementary analyses, and any supplemental risk assessment. The prior PHA, MOC forms, and any completed checklists or worksheets are appended to the cover report.

With *Update* documentation, whether the *Update* was by detailed review or change and incident review, the revalidation team can, and should, facilitate future revalidations by keeping a record of what was changed in the revalidation. This can be accomplished by any of the following:

- Editing a copy of the prior PHA in the "Track Changes" or "Revalidation" mode of the PHA documentation software
- Using a different color ink (or font color) to indicate changed items
- Adding detailed annotation of the items that were changed in MOC summary sheets, incident review sheets, or in a "comments" column of the worksheets documented in the PHA

Revalidated complementary analyses or supplemental risk assessments can also be documented in the *Redo* or *Update* style. Update documentation can be used when the original analysis is in a format that is not easily annotated (e.g., a hard copy or scanned document). However, many companies will justify a *Redo* of the PHA (using *Redo* documentation) if they find issues with the previous documentation, such as documentation in a non-editable format.

An important distinction is that the *Redo* documentation style should always be used when the PHA is being *Redone*. *Update* documentation style is not acceptable for a full *Redo* of the PHA. However, if the PHA is being revalidated by an *Update*, either the *Update* or *Redo* documentation style can be used. Some companies have found that having a fully documented PHA after each revalidation (i.e., using the *Redo* documentation style as discussed previously, regardless of methodology) reduces the need to reference existing reports and helps to ensure accurate and up-to-date "baseline" documentation of the PHA.

When contracted third-party PHA facilitators are used and the PHA report is within their scope of work, a *Redo* report that describes what they specifically did is usually the work product that will be delivered. If an *Update* report is desired from the contractor, the facility usually must provide the previous PHA in an editable form.

8.2   REPORT AND ITS CONTENTS

Some companies have a specified format for the documentation of initial PHAs and subsequent revalidations. Absent that, the following is a possible table of contents for the revalidation report:

Report Body.
- Executive summary of the revalidation activities, the elevated risks identified, and the recommendations to reduce them
- Purpose and introduction
- Description of the scope of the revalidation (e.g., the process unit/sections studied and applicable requirements for inclusion in the PHA). Note: For clarity, this may mention out-of-scope items, particularly those analyzed in other studies.
- Study dates and member attendance
- Table of team members indicating their job function (e.g., operator, engineer) *and indicating those team members who meet specific requirements for PHA team expertise*
- Description of revalidation approach, including justification of how specific requirements were met
- Summary of recommendations
- Summary of loss scenarios with elevated risks deemed as low as reasonably practicable (ALARP)

Information Included in Appendices.
- List of P&IDs, including revision number or date of version used
- List of study sections or nodes
- Copy of P&IDs, showing study nodes
- List of PSI referenced, with copies of anything not version-controlled
- Summary of assumptions
- List of MOCs considered
- List of previous incidents considered

Study Worksheets and Checklists.
- Core analysis (e.g., HAZOP, What-If)
- Supplemental risk assessments (e.g., LOPA, bow tie)
- Change Summary Worksheet (See Appendix C for example)

- Checklist of Process, Facility, and Human Factors Changes (See Appendix D for example)
- Facility Siting Checklists (See Appendix E for example)
- Human Factors Checklists (See Appendix F for example)
- External Events Checklists (See Appendix G for example)

Other Documented Information, *if used.*

- Essential Criteria Checklist (See Appendix A for example)
- PHA Quality and Completeness Checklist (See Appendix B for example)
- Company-Specific PSI Adequacy Checklist
- Wrap-Up Discussion Checklist (See Table 7-2 for example.)

Additional information on documentation for various hazard evaluation techniques is provided in the CCPS books *Guidelines for Hazard Evaluation Procedures* [2] and *Guidelines for Process Safety Documentation* [41].

8.3   RECOMMENDATIONS AND FOLLOW-UP

Recommendations. During the PHA revalidation meetings, the team is responsible for judging and reaffirming current risks against the organization's risk criteria. The most valuable result of a revalidation is the team's conclusions related to risks that exceed the organization's current risk tolerance. In addition, the team may find some risks that are tolerable, but not ALARP. In either situation, the team offers its recommendations on how management might lower the risk.

After the team's submission of the revalidated PHA, management (and their designees) are responsible for resolving the recommendations. Thus, the team

---

Example Recommendation Lacking in Detail:

"Add an alarm for low cooling water flow to Reactor 1."

Example Recommendation with Improved Detail:

"Provide a means to stop energy input to Reactor 1 on loss of cooling water. The risk of a runaway reaction could be reduced, for example, by stopping steam flow or stopping monomer feed if either low cooling water flow (FI-123) or high Reactor 1 temperature (TI-125) is detected."

should write each recommendation so that it "stands alone" and can be understood by someone who did not participate in the revalidation meeting. Each recommendation should state the issue as specifically as possible, with reference to the particular piece(s) of equipment or procedure(s) involved and/or any applicable risk ranking, policy, recognized and generally accepted good engineering practices (RAGAGEP), regulation, or law that prompted the recommendation. Well-written recommendations often include a brief explanation of the team's concern and rationale. This is frequently followed by any specific risk-reduction ideas the team thinks would help the person assigned responsibility for completing the recommendation. However, it is not the revalidation team's responsibility to design the solution, as discussed in Section 7.2.3, Example 6.

Follow-Up. As with any PHA recommendation, the recommendations coming out of the revalidation study should be resolved in a timely manner [2, p. 281]. Experience indicates that explicit, clear documentation of responsibilities is essential to the successful resolution of recommendations. In general, organizations resolve recommendations in a two-phase process, and some resolutions require authorization from upper management. **Phase 1** is the decision/authorization phase. When management receives the recommendations, it should quickly make an initial assessment of them within the bounds of any regulatory or organizational constraints. Some recommendations may be rejected because: (1) management is willing to accept continued operations with elevated risk, (2) management concludes the risk is already ALARP, or (3) management disagrees with the revalidation team's rationale for making the recommendation. Note: The risk management procedures of some organizations require corporate management to review and concur with local management decisions.

Management should convene a meeting with the revalidation team to explain their preliminary decision to reject any recommendations and give the team an opportunity to justify or explain the recommendation. Regardless of whether a recommendation is declined or implemented, these decisions should be documented in the facility recommendation tracking system. Declined recommendations should not simply be removed from the revalidation report.

The recommendation implementation schedule is often set during this management meeting. Recommendation benefits (risk reduction), recommendation complexity, resource availability, and implementation opportunities (e.g., during a unit shutdown) should be considered when assigning the completion dates for a particular recommendation. The rationale for extended recommendation completion dates should be documented, along with any interim plans for risk mitigation.

Those recommendations supported by management then enter **Phase 2**, which is the implementation phase. Each authorized recommendation (or interrelated group thereof) essentially becomes a follow-up project with a budget and schedule (and should be accompanied by applicable MOC documentation).

Typical documentation associated with resolution of a recommendation includes:

- Initial project charter
  - o   What is to be done
  - o   Schedule for completion, with justification for the time required
  - o   Who is responsible
  - o   Frequency of status reports until implementation is complete
  - o   Resources/budget
- How the recommendation was implemented
- MOC and PSSR records
- Date the recommendation was implemented
- List of employees trained or informed of the change, as appropriate

Some organizations also include a record of any metrics or follow-up analysis showing the effectiveness of the change as input for the next revalidation.

Not every authorized recommendation is implemented or implemented in the exact manner it was originally envisioned. Sometimes an appealing idea to reduce one risk introduces other risks that are just as high, or worse. Sometimes the cost of implementation proves to be grossly disproportionate to the benefits. Sometimes an alternate, and preferred, resolution is developed. If a recommendation was initially authorized, but later rejected or altered for any reason, the records associated with that ultimate resolution must also be kept.

8.4   RECORDS RETENTION AND DISTRIBUTION

The current, revalidated PHA report is a valuable source of information for any PSM program needing information about what could go wrong, how loss events might occur, and what safeguards/independent protection layers are in place to detect, prevent, or mitigate such events. For example, in addition to serving as the baseline for future MOC reviews, these documents provide valuable scenarios for initial and refresher training activities. Companies often keep current electronic copies of their PHAs on their intranet to make this information conveniently accessible to all employees. When the PHA revalidation report is issued, an electronic link can be easily sent to everyone who works in that process area, those who might be affected by its hazards, and anyone else who needs to know the revised PHA is available.

For the benefit of the next revalidation team, the PHA files could include a record, such as an end-of-PHA checklist, identifying what information was used, how it was used, and where it is stored (if not with the PHA itself). If a third-party facilitator was used, a copy of their digital files should also be retained in the PHA files.

Finally, it is important that this valuable information be protected. Company policies or procedures often establish retention, redundancy, and diversity requirements for the PHA-related records. Regulations may impose additional requirements on some facilities. For example, in the United States, companies with processes covered by the OSHA PSM or EPA RMP regulations must keep the initial PHA and all subsequent PHA revalidation reports, as well as documentation of the resolution of recommendations from these reports, for the life of the process. Covered facilities should ensure that archival copies of all versions of the PHA documentation are maintained, regardless of the documentation option used for prior PHA reports.

The "life of the process" may span decades, so the archive copies should resist both physical loss (e.g., flood, fire, theft) and technological obsolescence (e.g., outdated software, inaccessible storage media). When a process unit is dismantled or demolished, its productive "life" has ended, and the PHA records can be discarded or retained in accordance with company policy. However, if the process is merely shut down, decommissioned, or "mothballed," (even if they have no plans to resume production), it is prudent to retain the PHA records. Circumstances change, and years later there may unexpectedly be a need to resume production. If the PHA records were kept, revalidation would be an option; if not, an entirely new PHA would have to be completed before production could resume. Even in cases where the process equipment is ultimately dismantled and removed, access to PHAs could be of value to anyone evaluating the hazards of remediation efforts, which might be needed to allow new construction at the site.

8.5 PRINCIPLES FOR SUCCESSFUL DOCUMENTATION AND FOLLOW-UP

Actual experience in conducting revalidations, by PHA practitioners across a number of companies, has highlighted some keys for success, as well as some things that can impede success. While none of the items listed are absolute rules, they do provide valuable guidance.

Successful Practices:

- Documenting the revalidation to maximize its value to future users
- Documenting the PHA in a way that simplifies future revalidations (e.g., facilitating the *Update* approach)
- Describing complete loss scenarios, from initiating cause, through intermediate events, to potential consequences
- Documenting the changes made in an *Update* clearly (e.g., by using a distinct font color, using software features to track changes, or making detailed annotations in a comment column)
- Consolidating the core, complementary, and or supplemental analyses, along with any relevant portions of prior PHAs in a single revalidated PHA document that satisfies both regulatory and organizational requirements
- Writing recommendations that are clear to people who did not participate in the revalidation
- Minimizing the time between the PHA team developing recommendations and communicating them to management
- Resolving recommendations as soon as possible/practical
- Retaining a complete set of records for the next revalidation

Obstacles to Success:

- Documenting team discussions with abbreviated or cryptic notes, or using terminology unfamiliar to operating personnel
- Leaving blanks in worksheets, tables, or checklists
- Having no clear rationale for *Update* vs. *Redo* documentation style
- Documenting the PHA in *Update* style when a *Redo* is performed
- Using *Update* documentation on many sequential revalidations
- Failing to revise all the core, complementary, and supplemental study documentation affected by the revalidation

- Failing to quickly resolve any pending recommendations from prior PHAs that are reaffirmed by the revalidation team
- Failing to maintain archive copies of PHA documents in a format accessible to users as required (e.g., for the life of the process)
- Failing to notify affected personnel and contractors of PHA document location and recommendation resolution
- Receiving inadequate site or corporate leadership support for resolving the recommendations resulting from a PHA revalidation

REFERENCES

[1] Center for Chemical Process Safety, "CCPS Process Safety Glossary," 2019. [Online]. Available: www.aiche.org/ccps/resources/glossary.

[2] CCPS, Guidelines for Hazard Evaluation Procedures, Third Edition, Hoboken, NJ, USA: John Wiley & Sons, 2008.

[3] CCPS, Guidelines for Risk Based Process Safety (RBPS), Hoboken, NJ, USA: John Wiley & Sons, 2007.

[4] CCPS, Guidelines for Process Safety During the Transient Operating Mode: Reducing the Risks During Start-ups and Shut-downs, Hoboken, NJ, USA: John Wiley & Sons, 2021.

[5] American Petroleum Institute, "Recommended Practice 752, Management of Hazards Associated with Location of Process Plant Permanent Buildings," Washington, DC, USA, 2009.

[6] American Petroleum Institute, "Recommended Practice 753, Management of Hazards Associated with Location of Process Plant Portable Buildings," Washington, DC, USA, 2007.

[7] American Petroleum Institute, "Recommended Practice 756, Management of Hazards Associated with Location of Process Plant Tents," Washington, DC, USA, 2014.

[8] Chemical Industry Association, Guidance for the Location and Design of Occupied Buildings on Chemical Manufacturing and Similar Major Hazard Sites, 4th ed., 2020.

[9] CCPS, Guidelines for Siting and Layout of Facilities, Second Edition, Hoboken, NJ, USA: John Wiley & Sons, 2018.

[10] CCPS, Guidelines for Preventing Human Error in Process Safety, Hoboken, NJ, USA: John Wiley & Sons, 2010.

[11] CCPS, Guidelines for Inherently Safer Chemical Processes: A Life Cycle Approach, Third Edition, Hoboken, NJ, USA: John Wiley & Sons, 2019.

[12] American Petroleum Institute, "Recommended Practice 571, Damage Mechanisms Affecting Fixed Equipment in the Refining Industry," Washington, DC, USA, 2020.

[13] American Petroleum Institute, "Recommended Practice 970, Corrosion Control Documents," Washington, DC, USA, 2017.

[14] CCPS, Guidelines for Combustible Dust Hazard Analysis, Hoboken, NJ, USA: John Wiley & Sons, 2017.

[15] National Fire Protection Association, "NFPA 652, Standard on the Fundamentals Combustible Dust," Quincy, MA, USA, 2019.

[16] CCPS, Layers of Protection Analysis: Simplified Process Risk Assessment, Hoboken, NJ, USA: John Wiley & Sons, 2001.

[17] CCPS, Guidelines for Initiating Events and Independent Protection Layers in Layers of Protection Analysis, Hoboken, NJ, USA: John Wiley & Sons, 2015.

[18] CCPS, Guidelines for Enabling Conditions and Conditional Modifiers in Layers of Protection Analysis, Hoboken, NJ, USA: John Wiley & Sons, 2014.

[19] CCPS and the Energy Institute, Bow Ties in Risk Management: A Concept Book for Process Safety, Hoboken, NJ, USA: John Wiley & Sons, 2018.

[20] CCPS, Guidelines for Process Equipment Reliability Data, with Data Tables, Hoboken, NJ, USA: John Wiley & Sons, 1989.

[21] CCPS, Guidelines for Chemical Process Quantitative Risk Analysis, Second Edition, Hoboken, NJ, USA: John Wiley & Sons, Inc., 2000.

[22] J. A. Klein, "Two Centuries of Process Safety at DuPont," *Process Safety Progress*, vol. 28, no. 2, pp. 114-122, 2009.

[23] T. Kletz, "ICI's Contribution to Process Safety," *Loss Prevention in the Process Industries, Symposium Series*, no. 155, pp. 39-42, 2009.

[24] T. Ennis, "On the Integration of Hazard & Risk Studies into a Project Environment," *Loss Prevention in the Process Industries, Symposium Series*, no. 156, pp. 47-52, 2011.

[25] CCPS, Incidents That Define Process Safety, Hoboken, NJ, USA: John Wiley & Sons, 2008.

[26] State of New Jersey, "Toxic Catastrophe Prevention Act Program, Consolidated Rule Document (Revision 10)," New Jersey Administrative Code 7:31-4.2, p. 51, Trenton, NJ, USA, 2016.

[27] State of California, "Process Safety Management for Petroleum Refineries," California Code of Regulations 5189.1(e)(3)(C), p. 5, Sacramento, CA, USA, 2017.

[28] International Institute of Ammonia Refrigeration, "Guidelines for IIAR Minimum Safety Criteria for a Safe Ammonia Refrigeration System (Bulletin No. 109)," Washington, DC, USA, 1997.

[29] International Electrotechnical Commission, "IEC 61508, Standard for Functional Safety of Electrical/Electronic/ Programmable Electronic Safety-Related Systems," Geneva, Switzerland, 2010.

[30] National Fire Protection Association,, "ANSI/NFPA 70, National Electric Code," Quincy, MA, USA, 2020.

[31] OSHA, "Akzo-Nobel Chemicals - Limits of a Process," 28 February 1997. [Online]. Available: https://www.osha.gov.

[32] CCPS Monograph, "Assessment of and Planning for Natural Hazards, Third Edition," New York, NY, USA, 2019.

[33] J. B. Babcock and W. M. Bradshaw, "The Right People – Key to a Successful Hazard Review," in *AIChE Spring National Meeting Proceedings*, Atlanta, GA, USA, 2005.

[34] U.S. Chemical Safety and Hazard Investigation Board, BP America Refinery Explosion, Washington, DC, USA, 2007.

[35] British Broadcasting Corporation, "On This Day 1 November 1986," 1 November 1986. [Online]. Available: www.news.bbc.co.uk.

[36] T. Kletz, An Engineer's View of Human Error, Third Edition, Boca Raton, FL, USA: CRC Press, 2001.

[37] CCPS, Guidelines for Safe Automation of Chemical Processes, Second Edition, Hoboken, NJ, USA: John Wiley & Sons, 2017.

[38] CCPS, Guidelines for the Management of Change for Process Safety, Hoboken, NJ, USA: John Wiley & Sons, 2008.

[39] D. K. Whittle and K. Smith, "Six Steps to Effectively Update and Revalidate PHAs," Chemical Engineering Progress, pp. 70-77, January 2001.

[40] CCPS, Guidelines for Managing Process Safety Risks During Organizational Change, Hoboken, NJ, USA: John Wiley & Sons, 2013.

[41] CCPS, Guidelines for Process Safety Documentation, Hoboken, NJ, USA: John Wiley & Sons, 1995.

[42] CCPS, Guidelines for Developing Quantitative Safety Risk Criteria, Hoboken, NJ, USA: John Wiley & Sons, 2009.

APPENDICES

Electronic, editable versions of the checklists in these appendices are available online at www.aiche.org/ccps/publications/reval2ed

Password: revalbooked2

APPENDIX A ESSENTIAL CRITERIA CHECKLIST*

OBJECTIVE: To evaluate the prior PHA against essential criteria established by company and regulatory requirements. A "No" answer to any of the following questions strongly indicates a *Redo* of the PHA is warranted.

---

* This checklist is provided for illustrative purposes only and is specific to United States regulations. Readers may wish to develop such a checklist specific to their own situation, requirements, and needs.

CRITERIA	Yes/No
Prior PHA Methodology *(Section 3.1.1)*	
Was the PHA method used from this approved list or was an appropriate equivalent methodology used? • What-If • Checklist • What-If/Checklist • Hazard and Operability (HAZOP) Study • Failure Mode and Effects Analysis (FMEA) • Fault Tree Analysis (FTA)	
Was the prescribed PHA method appropriate for the complexity of the process (i.e., was it suitably rigorous for identifying the process hazards present and evaluating their risk)?	
Was the prescribed PHA method applied in accordance with company guidelines?	
Prior PHA Inputs *(Section 3.1.2)*	
Did the qualifications of the PHA leader/facilitator meet all company and regulatory requirements (i.e., was the leader/facilitator trained, experienced, and competent in the PHA method used)?	
Did the PHA team make-up/qualifications meet all company and regulatory requirements include, as a minimum: • Operations team member(s) with adequate process and equipment knowledge, including recent hands-on operating experience? • An engineer with industry and specific process knowledge and experience? • An employee who has experience and knowledge specific to the process being evaluated?	
Was the PHA based on current, complete, and accurate process safety information (PSI)?	
Prior PHA Scope *(Section 3.1.3)*	
Did the physical scope include all equipment "covered" by regulation and company policy?	
Did the PHA address all the relevant operational configurations (e.g., did the team consider different batch operations, seasonal	

CRITERIA	Yes/No
operations, startup, cleanout, maintenance, and shutdown configurations)?	
Prior PHA Documentation *(Section 3.2 and Chapter 8)*	
Is the PHA documentation sufficient, or can sufficient documentation be reconstructed to: • Indicate PHA team meeting dates? • Verify the five-year history of process safety incidents, and any others with the potential for catastrophic consequences, were reviewed by the PHA team? • Verify the PSI used by the team was current and adequate to ensure a thorough study? • Identify the hazards, engineering and administrative controls (safeguards), and consequences if those risk controls fail? • Verify that facility siting was addressed? • Verify that human factors were addressed? • Verify that a range of the possible safety and health effects was evaluated (e.g., by risk ranking or some other documented technique)? • Verify compliance with any additional regulatory or internal requirements? *Note: if only a few of these bullets are an issue, they are probably fixable via an Update; however, if several of them are an issue, then there are sufficient deficiencies to warrant a "No" answer.*	

APPENDIX B PHA QUALITY AND COMPLETENESS CHECKLIST*

OBJECTIVE: To evaluate the prior PHA against quality and completeness criteria established by company and regulatory requirements. A "No" response to any item requires that the issue be adequately addressed during the PHA revalidation.

The columns in this example checklist are generic **Q**, **T**, and **E**.

- **Q** - Question column. Typically, questions are organized by topic. The writers of this book expect that when used in industry, these questions can be modified and updated by facilities to meet their needs and unique situations. For example, the topics listed in these checklists could be used (with or without the full listing of questions) to help a team consider the quality of a prior PHA.

- **T** - Team evaluation column. The team should enter their response to the question (e.g., "Yes," "No," or "Not Applicable"), followed by brief documentation of the discussions and justification of the response. While it is generally acceptable for occasional responses to contain a simple "Yes" or "No," more responses should contain some detail of the team discussions, justifications, and concerns (if any). If safeguards or controls protect against the topic of the question, those should also be listed in this, or a separate, column.

- **E** - Evidence of compliance/Team comments column. Brief documentation of the discussions and evidence supporting the team evaluation. Even if the evidence seems obvious, it is generally better to document some detail of the team discussions, justifications, and concerns (if any).

---

* This checklist is provided for illustrative purposes only and is specific to United States regulations. Readers may wish to develop such a checklist specific to their own situation, requirements, and needs.

Q	T	E
Application of Analysis Methodology *(Section 3.2.1)*		
Is the PHA method applied consistently throughout the prior study? Evidence includes: • A description of the PHA method used • Proper documentation of an *Update*, such as "No new causes discovered" or "No new issues" when the team could not identify additional unique causes for a deviation		
If the What-If method was used: • Were the analysis nodes small enough to identify hazards? • Was a checklist used to formulate questions? • Were all appropriate questions documented for each node? • Were most questions about hazards for which the safeguards were deemed adequate?		
If the Failure Modes and Effects Analysis (FMEA) method was used: • Was each component boundary clearly defined? • Were concurrent safeguard failures analyzed? • Were reactive chemical hazards considered? • Were human errors considered?		
If the Hazard and Operability (HAZOP) Study method was used: • Were the analysis nodes small enough to identify hazards? • Were all applicable deviations in each node documented, even if there were no consequences of interest? • Were deviation meanings defined consistently? • Were loss of containment deviations considered?		
Is there evidence that required process safety information (PSI) was available and up to date (or recommendations were made to complete the PSI)? Such as: • Safety data sheets (SDSs) • Process chemistry • Maximum intended inventory • Operating limits • Equipment information (e.g., design temperatures/pressures) • Piping and instrumentation diagrams (P&IDs) • Electrical classification • Relief system design and sizing • Ventilation design • Codes and standards • Material and energy balances • Safety system design		

Q	T	E
Does the PHA report indicate all modes of operation were considered using questions, guidewords, and failure modes? Is there documentation to support that the PHA addressed the unit hazards during non-routine operations (e.g., startup, shutdown, emergency, maintenance, sampling)?		
For revalidated PHAs, is there documentation of the last change included in the prior PHA scope?		
If checklists or What-If/Checklists were used, is there evidence that: • Questions were answered as intended (no blank or simple "Yes/No" answers)? • Answers are sufficiently detailed? • Responses make technical sense? • Changes to existing recognized and generally accepted good engineering practices (RAGAGEP) were considered?		
For initial studies where the team did not have actual operating experience with the process or a similar process (e.g., projects for de-bottlenecking or major modifications), are those nodes for which the team lacked this experience currently operating using the design limits assumed in the initial PHA?		
Did the PHA team review/discuss previous incidents associated with the unit? Evidence may include: • A list or copy of incident reports • A worksheet documenting incident discussions		
Were near-miss incidents included in discussion of previous incidents?		
If the HAZOP or FMEA method was used, was there evidence that nodes were well defined and deviations/failure modes had meaning?		
Were causes meaningful and sufficiently detailed?		
Did the PHA consider equipment failures (fatigue, vibration, defective material, overpressure, structural support failure, weld failure, corrosion, erosion, etc.) as causes of accidents even though the equipment may have safety or shutdown systems specifically designed to prevent the postulated failure?		

Q	T	E
Is there evidence that the PHA considered safeguards and controls such as: • Relief protection? • Isolation valves? • Emergency shutdown systems? • Release detection systems (e.g., hydrocarbon or toxic gas monitors)? • Secondary containment and dikes around large liquid inventories? • Flares, scrubbers, and release mitigation systems? • Fireproofing and fire suppression systems? • Personal protective equipment (PPE) (self-contained breathing apparatus [SCBA], protective clothing, showers, respirators, etc.)? • Emergency response plans? • Equipment redundancy (pumps, instruments, backup power supplies for critical components, etc.)?		
Did the PHA report document both engineering and administrative risk controls (e.g., safeguards or defenses in PHA worksheets, listing of general protections) applicable to the process hazards identified?		
Did the PHA consider process instrumentation, alarms, interlocks, programmable logic controllers (PLCs), distributed control system (DCS), etc.?		
Did the PHA consider maintenance/testing of equipment, instrumentation/alarms, etc.?		
Is there evidence that human factors issues were addressed by: • Performing a separate human factors checklist analysis? • Performing a walkthrough of the process unit or reviewing a plot plan of the facility/unit? • Performing a review of procedures with a focus on reducing the potential for human errors (ambiguous steps, missing steps, wrong steps, etc.)? • Incorporating human errors as causes of potential loss scenarios identified and evaluated by the team? • Identifying labeling or other ergonomic issues? • Identifying operator staffing/workload/schedule issues? • Making recommendations addressing ways to reduce human errors?		

Q	T	E
Is there evidence that specific siting issues were addressed by: • Performing a separate siting checklist analysis? • Performing a walkthrough of the process unit or reviewing a plot plan of the facility/unit? • Reviewing a separate facility siting study, considering issues such as building occupancy, distance from flammables/toxics, design criteria, and potential consequences to both on-site and off-site persons caused by accidents in the unit? • Considering siting issues pertaining to location of buildings (separation distances, location of air intakes, ignition sources, drainage, etc.)?		
Were control room construction and applicable safeguards against releases considered (detectors, pressurization and alarms, intake elevation, blast resistance, separation distance, PPE, etc.)?		
Did the PHA consider electrical classification for the unit?		
Is there evidence that operating procedures were available and up to date?		
Were common cause failures of redundant equipment or personnel considered in the PHA?		
Did the PHA document the consequences, assuming that all engineering and administrative controls (e.g., relief valves, shutdown systems, alarms, trips, interlocks) fail when demanded to mitigate or stop a loss scenario?		
Are worst credible consequences documented?		
Is the documentation of consequence severity logical and consistent?		
If successful operation of safeguards (e.g., relief valve discharge to the atmosphere) could lead to a hazard, are these secondary consequences documented?		
Did the PHA focus on identifying reliable and independent safeguards?		
Did the PHA rely solely on safe work procedures or other administrative controls (e.g., operating procedures, lockout/tagout procedures, work permits, management of change [MOC]) to eliminate or reduce the severity of loss scenarios that could be serious if administrative controls are ignored or violated?		
Were the flammable, reactive, and toxic chemicals (including highly hazardous chemicals [HHCs]) documented in the PHA?		

Q	T	E
Were all pertinent hazards (fire, explosion, boiling liquid expanding vapor explosion [BLEVE], toxicity, chemical burn, asphyxiation, etc.) associated with releases addressed in the PHA?		
Was all equipment containing HHCs, or that could contain HHCs, addressed in the PHA? (Compare the analysis nodes to process flow diagrams and/or P&IDs)		
Was contamination of process chemicals addressed in the PHA?		
Was loss of utilities addressed in the PHA?		
Was the unit flare header (and knock out drum if one is provided) addressed in the PHA or in a separate PHA?		
Were the documented hazards consistent with hazards listed on safety data sheets (SDSs) for HHCs?		
Have incidents since the prior PHA revealed new hazards that were not previously identified?		
Was there evidence that the PHA team evaluated the range of effects for loss scenarios postulated (e.g., impact area, references to SDSs, health/safety effects possibly identified as part of the hazards of the process, use of a risk matrix)?		
Was there evidence the organization's risk tolerance criteria were consistently and correctly applied?		
Were there recommendations for additional risk controls where consequences of interest were listed in the worksheets with no (or weak) existing safeguards?		
If the prior PHA included recommendations for operability improvements, were their resolutions documented? Were any safety implications of the economic loss scenarios overlooked?		
Did the documentation for closure of previous safety recommendations meet all requirements?		

APPENDIX C EXAMPLE CHANGE SUMMARY WORKSHEET

OBJECTIVE: To record both controlled and uncontrolled changes identified during documentation reviews and interviews. Changes will be reviewed and addressed during PHA revalidation sessions.

Drawing/ Document Number	Identified Change	Source	Reference/ Comments	Action for Revalidation

APPENDIX D CHECKLIST OF PROCESS, FACILITY, AND HUMAN FACTORS CHANGES*

OBJECTIVE: To serve as an aid in identifying and recording changes that may have occurred in the process, facility, or human factors. Changes will be reviewed during PHA revalidation sessions.

Topic/Question	Y/N	MOC No.
Process Changes		
What was the last MOC implemented before or in conjunction with the completion of the previous PHA?		
Is safety and health assessment or hazard evaluation documentation available for process modifications made since the previous PHA?		
Have process modifications been evaluated to determine whether additional engineering or administrative controls are necessary to maintain continued safe operation of the process?		
Have process modifications been evaluated to determine whether additional emergency response capabilities or activities are necessary?		
Have process modifications been evaluated to determine whether engineering or administrative safeguards have been rendered ineffective (relief capacities exceeded due to higher unit throughputs, etc.)?		
Have all PHA issues (e.g., siting, human factors) been addressed for equipment added to the process?		
Have any utilities been changed (e.g., added, removed, larger, smaller, different properties)?		
Have safety system connections (e.g., nitrogen purge) been removed from process equipment?		
Have equipment relief valves been changed from atmospheric discharge to a closed system (or vice versa)?		
Are new hazards present due to new operating modes?		
Are new hazards present due to new process chemistry?		
Are new hazards present due to new chemicals being used?		

---

* This checklist was adapted from Whittle and Smith [39] and is provided for illustrative purposes only. Readers may wish to develop such a checklist specific to their own situation and needs.

Topic/Question	Y/N	MOC No.
Have the consequences of a release changed due to significantly modified operating conditions (e.g., temperature, pressure)?		
Have the hazards associated with equipment returned to service since the previous PHA been evaluated?		
Have the hazards associated with equipment removed from service since the previous PHA been evaluated?		
Are new recognized and generally accepted good engineering practices (RAGAGEPs) that apply to relief system design for equipment in this process being implemented or evaluated?		
Are new RAGAGEPs that apply to inspection and testing of equipment in the process being implemented or evaluated?		
Has the fuel changed for fired equipment (e.g., furnaces, reboilers)?		
Has the flue gas changed for fired equipment (e.g., routing, discharge point, backpressure, composition, dewpoint)?		
Have heating or cooling media been changed?		
Is a higher-pressure utility being used to prepare equipment for maintenance?		
Do equipment cleaning procedures require the use of a new chemical?		
Are new utilities being used to prepare equipment for maintenance?		
Have contaminant levels or contaminants in raw materials changed such that new hazardous deposits could be formed?		
Facility Changes		
Have any new units been built? How are they protected from fires or explosions in adjacent areas? Did the new unit create new fire or explosion hazards for existing units?		
Have buildings, trailers, temporary offices, contractor work areas, etc., been constructed or relocated since the previous PHA such that they could be affected by a release?		
Are there new or changed occupied structures off-plot or outside fence lines that could be affected by process hazards?		
Have staffing levels changed since the previous PHA such that capabilities to quickly respond to emergency situations have diminished?		
Have traffic patterns (e.g., new rail spur, new chemical truck routing) changed since the previous PHA? (e.g., Could the		

Topic/Question	Y/N	MOC No.
process be affected by a new release source? Can emergency responders reach the process easily? Are external impacts more likely?)		
Have there been any roadway closures or changes (e.g., security barriers) that might interfere with emergency vehicle access?		
Have there been any changes in traffic patterns that affect exit routes?		
Are there any significant new or larger inventories of hazardous materials?		
Have there been any changes in the locations of large inventories?		
Are there any hazardous materials now being received in larger containers (e.g., by tank truck instead of drums)?		
Have there been any changes in the location of material handling areas or material handling practices?		
Have atmospheric detectors for hazardous chemicals been relocated, removed, abandoned, or disabled?		
Are there new ignition sources near this process?		
Has the process sewer system changed (e.g., now a closed system, new streams in a common sump)?		
Have normal inventories of hazardous chemicals changed such that the consequences of a release have been significantly altered?		
Has spill containment equipment or systems changed?		
Have process water treatment systems changed?		
Are there new units or equipment nearby (inside or outside the facility fence line) that could affect this process?		
Has the area electrical classification changed such that some equipment is not properly rated for its service?		
Is new/modified equipment properly rated for its area electrical classification?		
Have there been any electrical classification changes near motor control centers (MCCs)?		
Have facility modifications increased the number of locations that should have personal protective equipment (PPE) available?		
Have facility modifications introduced a need to install or relocate existing safety shower/eyewash stations?		

Topic/Question	Y/N	MOC No.
Have personnel been relocated such that responding to equipment deficiencies (e.g., shutting down a pump with a leaking seal) may take a significantly longer time?		
Human Factors Changes		
Has the operator training program deteriorated since the previous PHA?		
Have maintenance practices deteriorated since the previous PHA?		
Have inspection and test practices deteriorated since the previous PHA, particularly for standby safety systems (e.g., interlocks, relief valves)?		
Have chains of command or delegation of authority for critical positions deteriorated?		
Have control displays changed such that information necessary to diagnose and respond to upset conditions is not readily accessible?		
Have operator communication systems changed?		
Are equipment and piping system labels being maintained?		
Are color coding systems for piping and components being maintained?		
Have emergency alarm or notification systems changed?		
Have safe work practice, such as those listed, deteriorated since the previous PHA? • Confined space entry • Entry into process area • Lockout/tagout • Line breaking • Lifting over process equipment		
Have fire detection or suppression systems been modified such that they require a different operator response?		
Have facility modifications made alarms difficult to see or hear?		
Have facility modifications resulted in the potential to overwhelm an operator with alarms during upset conditions?		

APPENDIX E EXAMPLE FACILITY SITING CHECKLISTS

Two example facility siting checklists are contained in this appendix. The first is based on the original checklist from the first edition of this book, and the second is a more streamlined checklist. No universal facility siting checklist is pertinent for all facilities or processes, and individual facilities should tailor these checklists to their individual and specific needs and processes.

Therefore, the columns in the example checklists in this appendix are generic **Q**, **T**, and **R**.

- **Q** - Question column. Typically, questions are organized by topic. The writers of this book expect that when used in industry, these questions can be modified and updated by facilities to meet their needs and unique situations. For example, the topics listed in these checklists could be used (with or without the full listing of questions) to help a team consider facility siting issues associated with a project change or for an *Update* of a process with minimal changes. Some questions may need to be repeated to evaluate facility siting risks thoroughly for different parts of a unit.

- **T** - Team evaluation column. The team should enter their response to the question (e.g., "Yes," "No," or "Not Applicable"), followed by brief documentation of the discussions and justification of the response. While it is generally acceptable for occasional responses to contain a simple "Yes" or "No," more responses should contain some detail of the team discussions, justifications, and concerns (if any). If safeguards or controls protect against the topic of the question, those should also be listed in this, or a separate, column.

- **R** - Recommendation column. If the team judges risk to be elevated, recommendation(s) should be made in this column. Some facilities require recommendations resulting from PHA checklists to be process safety in nature, while others are open to any recommendation, be it process safety, occupational safety, operational concerns, or otherwise.

Example Facility Siting Checklist 1

OBJECTIVE: To aid in identifying facility siting issues relevant to process safety issues within the scope of the PHA. The PHA team completing this checklist should seek assistance from subject matter experts as necessary.

Q	T	R
I. Spacing Between Process Components		
Are operating units and the equipment within units spaced and oriented to minimize potential damage from fires or explosions in adjacent areas?		
Are there multiple safe exit routes from each unit?		
Has equipment been adequately spaced and located to safely permit anticipated maintenance (e.g., pulling heat exchanger bundles, dumping catalyst, lifting with cranes) and hot work?		
Are vessels containing highly hazardous chemicals (HHC) located sufficiently far apart? If not, what hazards are introduced?		
Is there adequate access for emergency vehicles (e.g., fire trucks)?		
Can adjacent equipment or facilities withstand the overpressure generated by potential explosions?		
Can adjacent equipment and facilities (e.g., support structures) withstand flame impingement or radiant heat exposures?		
When provisions have been made for relieving explosions in process components, are the vents directed away from personnel and equipment locations?		
II. Location of Large Inventories		
Are large inventories of HHCs located away from the process area?		
Is temporary storage provided for raw materials and for finished products at appropriate locations?		
Are the inventories of HHCs held to a minimum?		
Where applicable, are reflux tanks, surge drums, and run-down tanks located in a way that avoids the concentration of large volumes of HHCs in any one area?		
Where applicable, has special consideration been given to the storage and transportation of explosives?		
Have the following been considered in the location of material handling areas: • Fire hazards? • Location relative to important buildings and off-site exposures? • Safety devices (e.g., sprinklers)? • Slope and drainage of the area?		

Q	T	R
III. Location of the Motor Control Center		
Is the motor control center (MCC) located so that it is easily accessible to operators?		
Are circuit breakers easy to identify?		
Can operators/electricians safely open circuit breakers? Have they been trained?		
Is the MCC designed such that it could not be an ignition source? Are the doors always closed? Is a no-smoking policy strictly enforced?		
Is the MCC being improperly used for shelter?		
IV. Location and Construction of Occupied Structures and Control Rooms		
Is the occupied structure built to satisfy current company overpressure and safe-haven standards?		
Does the construction basis satisfy acceptable criteria (e.g., the Factory Mutual recommendations)?		
Are workers in occupied structures (or escape routes from them) protected from the following: • Toxic, corrosive, or flammable sprays, fumes, mists, or vapors? • Thermal radiation from fires (including flares)? • Overpressure and projectiles from explosions? • Contamination from spills or runoff? • Noise? • Contamination of utilities (e.g., breathing air)? • Transport of hazardous materials from other sites? • Possibility of long-term exposure to low concentrations of process material (e.g., benzene)? • Impacts (e.g., from a forklift)? • Flooding (e.g., ruptured storage tank)?		
Are vessels containing HHCs located sufficiently far from occupied structures?		
Were the following characteristics considered when the occupied structure's location was determined: • Types of construction of the room? • Types/quantities of materials? • Direction and velocity of prevailing winds? • Types of reactions and processes? • Operating pressures and temperatures? • Fire protection? • Drainage?		
If windows are installed, has proper consideration been given to glazing hazards?		

Q	T	R
Is at least one exit located in a direction away from the process area? Do exit doors open outward? Are emergency exits provided for multi-story control buildings?		
Are the ends of horizontal vessels facing away from control rooms?		
Are critical pieces of equipment in the occupied structure well protected? Is barricading against impact adequate?		
Are open pits, trenches, or other pockets where inert, toxic, or flammable vapors could collect located away from control buildings or equipment handling flammable fluids?		
Where piping, wiring, and conduit enter the building, is the building sealed at the point of entry? Have other potential leakage points into the building been adequately sealed?		
Is the occupied structure located a sufficient distance from sources of excessive vibration?		
Is a positive pressure maintained in occupied structures located in hazardous areas?		
Could any structures fall on the occupied structure in an accident?		
Is the roof of the occupied structure properly supported for any heavy equipment or machinery?		
V. Location of Machine Shops, Welding Shops, Electrical Substations, Roads, Rail Spurs, and Other Likely Ignition Sources		
Are likely ignition sources (e.g., maintenance shops, roads, rail spurs) located away from release points for volatile substances (both liquid and vapor)?		
Are process sewers located away from likely sources of ignition?		
Are all vessels containing HHCs or components containing material above its flash point located away from likely sources of ignition?		
Are the flare and fired heater systems located to minimize hazards to personnel and equipment, with consideration given to normal wind direction and wind velocity, as well as heat potential?		
VI. Location of Engineering, Lab, Administration, or Other Occupied Buildings		
Have site buildings been screened to identify occupied buildings requiring further review?		
Are occupied buildings located away from inventories of HHCs?		
Are occupied buildings located away from release points for HHCs?		
Are workers in occupied buildings protected from the following: • Toxic, corrosive, or flammable sprays, fumes, mists, or vapors? • Thermal radiation from fires (including flares)? • Overpressure and projectiles from explosions? • Discharges from pressure relief vents?		

Q	T	R
• Contamination of utilities (e.g., water)? • Contamination from spills or runoff? • Noise? • Transport of hazardous materials from other sites? • Flooding (e.g., ruptured storage tank)?		
VII. Unit Layout		
Are large inventories or release points for HHCs located away from vehicular traffic within the plant?		
Could specific siting hazards be posed to the site from credible external forces such as high winds, earthquakes and other earth movement, utility failure from outside sources, flooding, natural fires, ice accumulation, and fog?		
Is there adequate access for emergency vehicles (e.g., fire trucks)? Are access roads free of the possibility of being blocked by trains, highway congestion, spotting of rail cars, etc.?		
Are access roads well engineered to avoid sharp curves? Are traffic signs provided?		
Is vehicular traffic appropriately restricted from areas where pedestrians could be injured or equipment damaged?		
Are cooling towers located such that fog that is generated by them will not be a hazard?		
Are the ends of horizontal vessels facing away from personnel areas?		
Is hydrocarbon-handling equipment located outdoors?		
Are pipe bridges located such that they are not over equipment, including occupied structures and administration buildings?		
Is piping design adequate to withstand potential liquid load?		
VIII. Location of the Unit Relative to On-site and Off-site Surroundings		
Is a system in place to notify neighboring units, facilities, and residents if a release occurs?		
Are the detection systems and/or alarms in place to assist in warning neighboring units, facilities, and residents if a release occurs?		
Do neighbors (including units, facilities, and residents) know how to respond when notified of a release? Do they know how to shelter-in-place and when to evacuate?		
Are large inventories or release points for HHCs located away from publicly accessible roads?		
Is the unit, or can the unit be, located to minimize the need for off-site or intra-site transportation of hazardous materials?		
Are workers in this unit protected from the effects of adjacent units or facilities (and vice versa), and are environmental receptors and the public also protected from the following:		

Q	T	R
• Releases of HHCs? • Toxic, corrosive, or flammable sprays, fumes, mists, or vapors? • Thermal radiation from fires (including flares)? • Overpressure from explosions? • Contamination from spills or runoff? • Noise? • Contamination of utilities (e.g., sewers)? • Transport of hazardous materials from other sites? • Impacts (e.g., airplane crashes, derailments)? • Flooding (e.g., ruptured storage tank)?		
IX. Location of Firewater Mains and Backup (e.g., Diesel) Pumps		
Are firewater mains easily accessible?		
Are firewater mains and pumps protected from overpressure and blast debris impact?		
Is an adequate supply of water available for firefighting?		
Are the firehouse doors pointed away from the process area so the doors will not be damaged by overpressure from an explosion?		
X. Location and Adequacy of Drains, Spill Basins, Dikes, and Sewers		
Are spill containments sloped away from process inventories and potential sources of fire?		
Have precautions been taken to avoid open ditches, pits, sumps, or pockets where inert, toxic, or flammable vapors could collect?		
Are process sewers that transport hydrocarbon closed systems?		
Are concrete bulkheads, barricades, or berms installed to protect personnel and adjacent equipment from explosion and/or fire hazards?		
Are vehicle barriers installed to prevent impact to critical equipment adjacent to high traffic areas?		
Do drains empty to areas where material will not pool and pose an elevated risk?		
Can dikes hold at least 110% of the capacity of the largest tank?		
Is there a means of access in and out of dikes, pits, etc.?		
XI. Location of Emergency Stations (Showers, Respirators, Personal Protective Equipment, etc.)		
Are emergency stations easily accessible?		
Is emergency response equipment and personal protective equipment (PPE) accessible both inside outside the zone of impact?		
Are first-aid stations prudently located and adequately equipped?		
Are safety showers heated/freeze protected/wind protected?		
Is there a control room alarm for water flow to safety showers and eyewash stations? (Is there a need for such an alarm?)		

Q	T	R
XII. Electrical Classification		
Is there an electrical classification document?		
Does the electrical classification appear correct and complete?		
Has the electrical classification document been recently revised?		
Have significant changes made since the system was originally constructed (addition of new materials, new sources of flammable gases or vapors, new low points [e.g., sumps or trenches] at grade) been included in the electrical classification document?		
Are the design and maintenance of ventilation systems adequate?		
Are there safeguards to alert operators when a ventilation system fails?		
Are ventilation systems being properly maintained, and are alarms and interlocks on these systems periodically function checked?		
Is adequate maintenance being done to function check natural ventilation systems?		
Are there technical bases for design changes to the ventilation systems?		
Are ventilation systems verified to be adequate for new gas or vapor loads?		
Are there adequate controls to ensure that electrically classified equipment is replaced with equipment of equal or higher classification?		
Are boundaries between electrically classified areas physical boundaries?		
Are Division 1 areas necessary (if there are any)?		
Are there adequate controls (e.g., a hot work permit system) on repair and construction activities, including work by contractor personnel?		
Are there specific requirements for personnel in classified areas (e.g., static dissipative attire, prohibition of ignition sources such as cell phones)?		
Does the electrical classification adequately reflect the effects of different modes of operation (e.g., normal operation, maintenance, startup, infrequent operating modes such as reactor regeneration or operation with a portion of the system bypassed)?		
XIII. Contingency Planning		
What expansion or modification plans are there for the facility?		
Can the unit be built and maintained without lifting heavy items over operating equipment and piping?		

Q	T	R
Are calculations, charts, and other documents available that verify siting has been considered in the layout of the unit? Do these siting documents show that consideration has been given to: • Normal direction and velocity of wind? • Atmospheric dispersion of gases and vapors? • Estimated radiant heat intensity that might exist during a fire? • Estimated explosion overpressure?		
Are appropriate security safeguards in place (e.g., fences, guard stations)?		
Are gates located away from the public roadway so that the largest trucks can move completely off the roadway while waiting for the gates to be opened?		
Where applicable, are safeguards in place to protect high structures against low-flying aircraft?		
Are adequate safeguards in place to protect employees against exposure to excessive noise, considering the cumulative effect of equipment items located close together?		
Is adequate emergency lighting provided? Is there adequate redundant backup power for emergency lighting?		
Are procedures in place to restrict nonessential or untrained personnel from entering areas deemed hazardous?		
Are indoor safety control systems (e.g., sprinklers, fire walls) provided in buildings where personnel will frequently be located, such as control rooms and administrative buildings?		
Are evacuation plans (from buildings, units, etc.) adequate and accessible to personnel?		
Are evacuation drills routinely conducted?		

Example Facility Siting Checklist 2

OBJECTIVE: To aid in identifying facility siting issues relevant to process safety issues within the scope of the PHA. The PHA team completing this checklist should seek assistance from subject matter experts as necessary.

Q	T	R
I. Process/Unit Assessment		
Have the following areas been evaluated for potential explosion, fire, and toxic hazards in the overall facility siting study		
• Control rooms? (See Section II)		
• Operating units?		
• Equipment (including flare and fired heater systems)?		
• Tank farms?		
• Process sewers?		
• Motor control centers (MCCs)?		
• Maintenance shops and established areas for cutting and welding?		
• Transportation routes (roads, rail spurs, etc.) within the site?		
• Pipe bridges located over occupied structures?		
• Fire protection equipment (including firehouse)?		
• Other occupied buildings (i.e., offices, warehouses, laboratories, analyzer rooms)?		
Are there vents or relief devices that discharge to the atmosphere? If yes, evaluate the following items:		
• Are employees or equipment near the discharge point, including downwind locations (e.g., on the ground, on platforms, on tanks or towers, on roofs)?		
• Are occupied building or control room air intakes in the vicinity of release points?		
Has the potential for vapor or liquid accumulations been considered for inert, toxic, or flammable chemicals? The following areas should be considered:		
• Drains		
• Spill containment areas (e.g., dikes, berms, ponds, remote impoundments)		
• High points in buildings or canopies over lighter-than-air flammables		

Q	T	R
• Open pits, trenches, ditches, sumps, or other pockets where highly hazardous chemicals (HHCs) could collect		
• Basements, below-ground confined spaces		
Are the inventories of HHCs held to a minimum?		
Are large inventories or release points for HHCs located away from public roads and plant traffic?		
Can dikes hold at least 110% of the capacity of the largest tank?		
Is there a safe means of access and egress for dikes, pits, etc.?		
Are MCCs located to be accessible in an emergency, and are circuit breakers easy to identify?		
Can qualified personnel safely open circuit breakers? Have qualified personnel been trained and has proper PPE been provided for this task?		
Is vehicular traffic appropriately restricted from areas where pedestrians could be injured or equipment damaged?		
Are vehicle barriers installed to protect process safety critical equipment and piping adjacent to heavy traffic areas?		
Are the ends of horizontal vessels oriented away from personnel areas?		
Is plant entry/egress secured to prevent unauthorized entry?		
Is entry to the process area controlled?		
Is the equipment generally well instrumented with rationalized and effective process alarms?		
Are appropriate means provided to shut down major pieces of equipment remotely?		
Have combustible gas and toxic gas detectors been installed at strategic locations?		
Is general and emergency lighting sufficient for personnel safety?		
Are ventilation systems adequate? Consider the following items:		
• Are written procedures provided for any shelters or safe havens?		
• Is there a design basis document that is current?		
• Are there alarms on ventilation systems where necessary and are the alarms and interlocks included in the inspection, test, and preventive maintenance (ITPM) program?		
For new or existing facilities/units, have the electrical classification requirements been identified and are they consistent with the current classification of the location and adjacent areas?		

Q	T	R
Are occupied buildings and major pieces of equipment (where appropriate) protected by fixed water spray systems (sprinklers or deluge) and is there adequate firewater supply capacity and pressure?		
Is fire response equipment (fixed monitors, portable monitors, sprinklers, emergency vehicles, etc.) easily accessible and capable of reaching all major equipment?		
Are emergency eyewash stations, safety showers and fire extinguishers properly located and easily accessible?		
Are emergency eyewash and safety showers freeze and heat protected, as needed?		
Is water "tempered" to a safe temperature range, considering the potential for scalding or hypothermia?		
Are first-aid stations appropriately located and adequately equipped?		
If automatic shutdown systems do not exist, is there an operator who can monitor the status of the process at all times?		
Is adequate protection provided if employees (other than emergency responders) are required to remain in the area during an emergency to monitor process safety equipment or to effect a safe shutdown?		
Is there an adequate emergency response plan for the site? (Is there an external community notification procedure?)		
Have safe evacuation routes been established that account for prevailing wind conditions and are evacuation drills routinely conducted? Possibly two assembly areas at 90° or more apart so both not affected by wind-blown releases from the process		
Are building and control room exits appropriately located? Consider the following items:		
• Is at least one exit located in a direction away from the process area?		
• Are all walkways and aisles free of obstructions?		
• Do exit doors open outward, if required?		
• Are vestibules (double door "air lock") provided at shelter/safe haven entries when multiple personnel will need to enter at different times as an incident evolves?		
• Are all exits, staircases, and corridors free of storage and clutter?		
• Are emergency exits provided for multi-story control buildings?		
If employees have to travel through a toxic vapor cloud to evacuate, has portable respiratory protection been provided?		

Q	T	R
Can emergency alarms be heard in all areas of the process?		
Have all plant areas been evaluated for potential vibration issues?		
II. Control Room Evaluation (These are also relevant questions for any occupied building)		
Are there adequate emergency exits?		
Has the control room been evaluated for potential explosion, fire, and toxic hazards? Also consider structures and heavy equipment that could collapse on the control room.		
Is the fresh air intake capable of being shut down from inside the control room, if it is specified as a shelter-in-place location?		
Is it possible to seal doors, windows, etc., against infiltration and are all other penetrations adequately sealed, if the control room is specified as a shelter-in-place location?		
Is the control room free of windows or other openings (beyond doors) unless specially reinforced?		
III. Location of Process Units Relative to Surrounding Areas (Consider for each process unit)		
• Are systems in place to notify nearby units, facilities, and neighbors of a release?		
• Do the surrounding units, facilities, and neighbors know how to respond to alarms/alerts?		
• Are large inventories of HHCs located away from publicly accessible roads?		
• Are workers in the unit/process protected from the potential impacts of incidents in adjacent or nearby facilities? These may include:		
o Releases of HHCs		
o Overpressure from explosions		
o Thermal radiation from fire		

APPENDIX F EXAMPLE HUMAN FACTORS CHECKLISTS

Two example human factors checklists are contained in this appendix. The first is based on the original checklist from the first edition of this book, and the second is a more streamlined checklist. No universal human factors checklist is pertinent for all facilities or processes, and individual facilities should tailor these checklists to their individual and specific needs and processes.

Therefore, the columns in the example checklists in this appendix are generic **Q**, **T**, and **R**.

- **Q** - Question column. Typically, questions are organized by topic. When used in industry, these questions can be modified and updated by facilities to meet their needs and unique situations. For example, the topics listed in these checklists could be used (with or without the full listing of questions) to help a team consider human factors issues associated with a project change or for an *Update* of a process with minimal changes.

- **T** - Team evaluation column. The team should enter their response to the question (e.g., "Yes," "No," or "Not Applicable") followed by brief documentation of the discussions and justification of the response. While it is generally acceptable for occasional responses to contain a simple "Yes" or "No," more responses should contain some detail of the team discussions, justifications, and concerns (if any). If safeguards or controls protect against the topic of the question, those should also be listed in this, or a separate, column.

- **R** - Recommendation column. If the team judges risk to be elevated, recommendation(s) should be made in this column. Some facilities require recommendations resulting from PHA checklists to be process safety in nature, while others are open to any recommendation, be it process safety, occupational safety, operational concerns, or otherwise.

Example Human Factors Checklist 1

OBJECTIVE: To aid in identifying human factors issues relevant to process safety issues within the scope of the PHA. The PHA team completing this checklist should seek assistance from subject matter experts as necessary.

Q	T	R
I. Housekeeping and General Work Environment		
Are adequate signs posted near maintenance, cleanup, or staging areas to warn workers of special or unique hazards associated with the areas?		
Are adequate barriers erected to limit access to maintenance, cleanup, or staging areas?		
Are working areas generally clean?		
Are provisions in place to limit the time that a worker spends in an extremely hot or cold area?		
Is noise maintained at a tolerable level?		
Are alarms audible above background noise both inside occupied structures and in the process area?		
Is normal and emergency lighting sufficient for all area operations?		
Is there adequate backup power for emergency lighting?		
Is the general environment conducive to safe job performance?		
II. Accessibility/Availability of Controls and Equipment		
Are adequate supplies of protective gear readily available for routine and emergency use?		
Are workers able to perform both routine and emergency tasks safely while wearing protective equipment?		
Is emergency equipment accessible without presenting further hazards to personnel?		
Is communications equipment both for normal operations and emergency situations adequate and easily accessible?		
Would others quickly know if a worker were incapacitated in a process area?		
Are the right tools (including special tools) available and used when needed?		
Is the workplace arranged so that workers can maintain a good working posture while performing necessary movements to conduct routine tasks?		
Is access to all controls adequate?		

Q	T	R
Can operators/maintenance workers safely perform all required routine/emergency actions, considering the physical arrangement of equipment (e.g., access to equipment or proximity of tasks to rotating equipment, hot surfaces, hazardous discharge points)?		
Are valves that require urgent manual adjustments (e.g., emergency shutdown) easily identifiable, readily accessible, and outside the hazard zone (or remotely operable)?		
Are sampling points located appropriately so operators are not subjected to hazards such as hot surfaces, asphyxiation, or chemical exposure?		
III. Labeling		
Is all important equipment (vessels, pipes, valves, instruments, controls, etc.) legibly, accurately, and unambiguously labeled?		
Does the labeling program include components (e.g., small valves) that are mentioned in the procedures even if they are not assigned an equipment number?		
Do all vales have clear open/closed position labels?		
Has responsibility for maintaining and updating labels been assigned?		
Are emergency exit and response signs (including windsocks) adequately visible and easily understood?		
Are signs that warn workers of hazardous materials or conditions adequately visible and easily understood?		
IV. Feedback/Displays		
Is adequate information about normal and upset process conditions clearly displayed in the control room?		
Are the controls and displays arranged logically to match operators' expectations?		
Are the displays adequately visible from all relevant working positions?		
Do separate displays present similar information in a consistent manner?		
Are automatic safety features provided when a process upset requires rapid response?		
Are automatic safety features provided when a process upset may be difficult to diagnose due to complicated processing of various information?		
Are the alarms displayed by priority?		
Are critical safety alarms easily distinguishable from control alarms?		
Is an alarm summary permanently on display?		

Q	T	R
Are nuisance alarms corrected and redundant alarms eliminated as soon as practical to help prevent complacency toward alarms?		
Have charts, tables, or graphs been provided (or programmed into the computer) to reduce the need for operators to perform calculations as part of the operation?		
If operators are required to perform calculations, are critical calculations independently checked?		
Does the computer check that values entered by operators are within a valid range?		
Do the displays provide an adequate view of the entire process as well as essential details of individual systems?		
Do the displays give adequate feedback for all operational actions?		
Are instruments, displays, and controls promptly repaired after a malfunction?		
Do administrative controls or procedures govern when instruments, displays, or controls are deliberately disabled or bypassed and that govern their return to normal service at the appropriate time?		
Does a formal mechanism exist for correcting human factors deficiencies identified by the operators (e.g., modifications to the displays, controls, or equipment to better meet operators' needs)?		
V. Controls		
Is the layout of the consoles logical, consistent, and effective?		
Are the controls distinguishable, accessible, and easy to use?		
Do all controls meet standard expectations (color, direction of movement, etc.)?		
Do the control panel layouts reflect the functional aspects of the process or equipment?		
Does the control arrangement logically follow the normal sequence of operation?		
Can operators safely intervene in computer-controlled processes?		
Can process variables be adequately controlled with the existing equipment?		
Do operators believe that the control logic and interlocks are adequate?		
Does a dedicated emergency shutdown panel or button exist? If so, is it in an appropriate location and guarded against accidental activation?		
Are alarm setpoints, interlocks, and control logic secured against unauthorized access and change?		

Q	T	R
VI. Workload and Stress Factors		
Is the control room always occupied (i.e., assigned duties do not require the control room operator to be absent from the control room)?		
Are the number and frequency of manual adjustments required during normal and emergency operations limited so that operators can make the adjustments without a significant chance of mistakes as a result of overwork or stress?		
Is the number of manual adjustments during normal operations sufficient to avoid mistakes as a result of boredom		
Have the effects of shift duration and rotation been considered in establishing workloads?		
Is the number of extra hours an operator must work if their relief fails to show up sufficiently limited so that worker safety is not adversely affected?		
Is the number of hours an operator or maintenance worker must work during startup or turnarounds sufficiently limited so that worker safety is not adversely affected?		
Can additional operators (e.g., from other areas or from off-site) be called in quickly at any time to help during an emergency?		
Is the staffing level appropriate for all modes of operation (normal, emergency, etc.)?		
VII. Procedures		
Do written procedures exist for all operating phases (i.e., normal operations, temporary operations, emergency shutdown, emergency operation, normal shutdown, and startup following a turnaround or after an emergency shutdown)?		
Are safe operating limits documented, providing consequences of deviating from limits and actions to take when deviations occur?		
Are procedures current (i.e., are they revised when changes occur, are they periodically reviewed)?		
Do operators believe that the procedure format and language are easy to follow and understand?		
Are the procedures accurate (i.e., do they reflect the way in which the work is actually performed)?		
Is responsibility assigned for updating the procedures, distributing revisions of the procedures, and ensuring that workers are using current revisions of the procedures?		
Are temporary notes or instructions incorporated into revisions of written operating procedures as soon as practical?		
Do procedures address the personal protective equipment required when performing routine and/or non-routine tasks?		

Q	T	R
VIII. Training (Employees and Contractors)		
Are new employees trained in the hazards of the processes?		
Do operators and maintenance workers receive adequate training in safely performing their assigned tasks before they are allowed to work without direct supervision?		
Does operator and maintenance worker training include training in appropriate emergency response?		
Do operators practice emergency response while wearing emergency protective equipment?		
Do operators practice emergency response during extreme environmental conditions (e.g., at night or when it is very cold)?		
Are periodic emergency drills conducted?		
Are emergency drills witnessed by observers and critiqued?		
Does a periodic refresher training program exist?		
Is special or refresher training provided in preparation for an infrequently performed operation?		
When changes are made, are workers trained in the new operation, including an explanation of why the change was made and how worker safety can be affected by the change?		
Are operators and maintenance workers trained to request assistance when they believe they need it to safely perform a task?		
Are operators and maintenance workers trained to recognize and report near misses as part of the incident investigation program?		
Are operators trained to shut down the process when in doubt about whether it can continue to operate safely?		

Example Human Factors Checklist 2

OBJECTIVE: To aid in identifying human factors issues relevant to process safety issues within the scope of the PHA. The PHA team completing this checklist should seek assistance from subject matter experts as necessary.

Q	T	R
I. Stress/Fatigue/Staffing		
Is the staffing level, competency, and experience appropriate for all modes of operations (normal, emergency, etc.)?		
Can operators perform all manual adjustments required during normal and emergency operations (not an excessive number of adjustments required)?		
Have the effects of shift duration and rotation been considered in established workloads (consider startup and turnarounds as well)?		
Are process safety critical alarms and notifications sounded in a continuously staffed area or other appropriate manner?		
Are workers able to perform tasks in a step-by-step format in the allotted time to reduce the "panic" factor?		
Can additional operators (e.g., from other areas or from off-site) be called in quickly at any time to help during an emergency or special circumstances?		
II. Communications		
Are shift turnover communications sufficient to adequately communicate plant operating conditions (e.g., Is shift turnover formal and logged? Is it maintained in an accessible log? Is shift turnover conducted in the actual work location of the positions being turned over? Is the overlap between shifts sufficient?)		
Is communications equipment adequate (amount and type) for the number of persons or stations who must communicate with each other?		
Are communications systems susceptible to electromagnetic interference during any operating mode of the plant?		
Are communications required at periodic intervals so that injured or incapacitated operators can be identified?		
Are there protocols/safeguards for protection of lone workers on site by themselves?		
Where remote operations are in place, are communications and travel time to the operations adequate?		

Q	T	R
Are adequate systems in place to safely operate remote equipment (consider communications between Control Room operators and Field operators, operator response time for alarms/emergency actions, adequate displays of operating information in control room, etc.)?		
Can communications equipment be used while wearing emergency personal protective equipment (PPE)?		
III. General Controls		
Are graphics and controls on control stations and instruments (control room and field) easily seen, manipulated, and understood from relevant working positions (all processes shown, in logical order, etc.)?		
Do controls on the distributed control system (DCS) and in the field have adequate responses and response times (including emergency/shutdown safety responses)?		
Are logic and controls (field and DCS) for emergency shutdown or process safety critical equipment adequate for the plant?		
Do controls check entered values against allowed ranges where necessary (min/max stops on valves, maximum addition rates, etc.)?		
Have graphs, tables, etc. been put in place where needed to minimize calculations and other work by operators and other personnel?		
Do all necessary field controls have visual identification for their position (green/red for open/closed valves) or movement of direction (arrows for open/close)?		
Are gauge faces/markings clearly visible, not obscured by weathering/fading/clouding, etc.?		
Is there any critical operational information needed quickly by operators that is difficult or time consuming to find (e.g., total flow into a tower that is several screens down and is not obvious)?		
IV. DCS Controls		
Do the as-found settings for safe operating limits/interlocks/ shutdown conditions match the descriptions in control item summary documents?		
Is it easy to distinguish between types of controls and alarms (normal, shutdown) and their priorities? Is the alarm summary, if present, easy to understand?		
Is adequate information shown on the computers at one time to allow for proper operation and control of processes?		
Are alarms updated to meet the needs of different process steps? Can all nuisance alarms that would cause a safety hazard be silenced, if necessary? Could they be eliminated?		

Q	T	R
Is it easy to navigate between screens/displays and do the screens/displays update quickly enough to allow for operator action/intervention of controls?		
Are initiation cues to move from one step to another acceptable?		
If the controls implement a sequence of steps, is it always clear to the operator which step the process is in?		
Are DCS lights and switches/controls logically arranged? Are colors consistent (e.g., red/green meaning stop/go or off/on OR red/green meaning danger/safe or energized/off)?		
V. Labeling		
Is process equipment legibly, accurately, and unambiguously labeled? (Including equipment necessary to operate the process and mentioned in operating procedures even if not assigned an equipment number; vessels, pipes, valves, instruments, controls, etc.)		
Do switch labels identify discrete positions (e.g., on/off, open/close)?		
Has responsibility for maintaining/ updating labels been assigned?		
Are signs (e.g., hazardous materials, colors not faded, not falling off, emergency exits, restricted entry, evacuation routes, windsocks) clearly visible? Are these signs easily understood (i.e., in a language the operators speak/read, considering location and condition)?		
Do labels correspond to drawings and procedures?		
VI. Procedures		
Do written procedures exist for all modes of operations, maintenance, and emergency response?		
Are the written operating procedures adequate (easy to understand and available)?		
Are the conditions warranting evacuation of the control room or site, and procedures to do so, documented?		
Are the procedures written in the language that the operators speak?		
Are the procedures written to match the estimated reading level of the workforce that will use the procedures?		
Are the actual procedures used in the operator training program, or are separate procedural training documents used?		
Do the procedures contain enough detail to control the process adequately (could a trained operator or technician return from a long absence and use the procedure safely)?		
Are warnings and cautions shown clearly with distinctive font and size in locations within the procedures where the user's attention will be drawn to them?		
Do the procedures contain drawings and other graphical information where necessary to clarify the required actions?		

Q	T	R
Are the procedures tabbed so that important sections, such as emergency operations, can be found quickly?		
Do the procedures clearly indicate what conditions require immediate shutdown of the process vs. allowing continuing efforts to bring the process back into normal control?		
Does the operator training program cover all the procedures and required operations?		
Is refresher training held periodically? How often?		
Is equipment clearly labeled and are the equipment tag and line numbers used in the procedures?		
Are the procedures readily accessible to the operators and supervisors during all shifts? During power outages?		
VII. Workplace and Personnel Issues		
Do untrained personnel work (operating or maintaining) in the unit?		
Are unit or control room operators overloaded during normal or emergency operations?		
What is the normal length of a work shift for control room operators? For in-plant operators?		
What is the maximum number hours an operator is allowed to work in a 24-hour period during normal operations or turnaround periods?		
Is overtime a normal part of any given operator's routine? How long is a typical overtime period?		
How many standards tasks or operations is an operator expected to carry out during a shift? How much time typically occurs between these tasks?		
Can operators become easily bored?		
Are control room operators required to leave the control room to perform assigned tasks?		
Has adequate PPE and emergency breathing equipment been provided for the expected personnel during emergency conditions?		
Can the operators carry out important operations while wearing PPE?		
Do the operators have the required vision, hearing, and strength to perform their required tasks?		
Do the ambient conditions (temperature, humidity, lighting, and noise) present strong impediments to performing required tasks safely, especially during emergency conditions?		
Do operators exhibit strong "cultural" mindsets or habits that may adversely affect the manner in which they do their jobs (e.g., taking shortcuts, modifying equipment for convenience, etc.)?		
Do collective bargaining negotiations or other labor disputes create operational problems?		

APPENDIX G EXAMPLE EXTERNAL EVENTS CHECKLIST

The columns in the example checklist in this appendix are generic **Q**, **T**, and **R**.

- **Q** - Question column. Typically, questions are organized by topic. When used in industry, these questions can be modified and updated by facilities to meet their needs and unique situations.

- **T** - Team evaluation column. The team should enter their response to the question (e.g., "Yes," "No," or "Not Applicable") followed by brief documentation of the discussions and justification of the response. While it is generally acceptable for occasional responses to contain a simple "Yes" or "No," more responses should contain some detail of the team discussions, justifications, and concerns (if any). If safeguards or controls protect against the topic of the question, those should also be listed in this, or a separate, column.

- **R** - Recommendation column. If the team judges risk to be elevated, recommendation(s) should be made in this column. Some facilities require recommendations resulting from PHA checklists to be process safety in nature, while others are open to any recommendation, be it process safety, occupational safety, operational concerns, or otherwise.

OBJECTIVE: To aid in identifying external events relevant to process safety issues within the scope of the PHA. The PHA team completing this checklist should seek assistance from subject matter experts as necessary.

Q	T	R
I. General		
Does the site have an emergency preparedness response plan that covers external events like natural disasters?		
Does the site's emergency preparedness response plan address all requirements of the corporate emergency response standard?		
Has the site clearly identified a communication and decision-making structure or plan for all process areas in case of a natural disaster?		
Does the site have adequate inventory of process safety critical materials (e.g., inhibitor, retardant, inert gas) if access to the site is interrupted?		
Does the site have access to emergency communication equipment (radios, satellite phones, etc.) in case of a natural		

Q	T	R
disaster or loss of local/regional telephone, internet, or cell networks?		
Does the site have access to emergency power (generators, trailers) in case of natural disaster or loss of off-site power? (Consider inventory such as temperature-sensitive peroxides.)		
For seasonally related potential natural disasters, does the site have plans to gather necessary supplies and/or to prepare prior to the upcoming season (e.g., for cold weather, completing a heat tracing check in early fall)?		
Has the site considered relevant geological events such as avalanches, landslides, subsidence, tsunamis, volcanic activity, or coastal erosion?		
If the site shares emergency responders, has the site accounted for lengthened emergency response in case of natural disaster?		
Does the site have a policy to verify operational readiness prior to startup after a natural disaster to ensure recommissioning is completed for a safe startup?		
Have plans been made to account for the security and preservation of key process safety information and procedures? <ul><li>Have hard copy files been stored in a location that is resistant to flood, wind, and other external impacts?</li><li>Have electronic files been backed up to a company server or otherwise preserved at another location away from the impact zone of the external hazard?</li></ul>		
II. Flooding (Consider rain events, hurricanes, tsunamis, dam damage, typhoons, seiches, tides)		
Has the credibility and periodicity of site flooding been evaluated (i.e., whether the site is in a flood plain)?		
If flooding is credible, has equipment location and height been evaluated against potential flood levels (emergency generators, storage tanks, etc.)?		
If flooding is credible, does the site have identified and documented actions to take if flooding occurs (e.g., powering down equipment, evacuation, inventory control, removal of hazardous inventory, removal of temperature-sensitive material, if applicable)?		

III. High Winds (Consider tornados, typhoons, and hurricanes)		
Does the site emergency preparedness plan cover whether hurricane-force wind is credible for the site?		
If hurricane-force wind is credible, does the site have documented design information regarding wind loading for fixed outdoor equipment (e.g., vent stacks, high structures)?		
If hurricane-force wind is credible, does the site have in its emergency plans what equipment needs to be moved or secured prior to high wind events?		
If high wind is credible, does the site have identified and documented actions to take if hurricane-force wind occurs? (e.g., shelter in place, powering down equipment)		
If tornados are credible for the site, does the site have a documented plan in case of tornado watch or tornado warning? (Consider emergency shutdown timing, shelter-in-place locations evaluated for effectiveness in tornado-strength winds, etc.)		
IV. Earthquake		
Does the site emergency preparedness plan cover whether earthquakes are credible for the site?		
If earthquakes are credible, does the site have documentation to determine if the site construction is designed for the magnitude of earthquake anticipated?		
If earthquakes are credible, has the site identified and documented actions to take if an earthquake occurs (e.g., powering down equipment, evacuation, shelter in place)?		
After an earthquake occurs, does the site have a post-earthquake checklist for safe startup?		
V. Cold Weather (Consider routine temperatures and also possible lows)		
If site experiences sub-freezing temperatures, is there appropriate heat tracing on outdoor lines?		
If site experiences sub-freezing temperatures, is heat tracing considered for equipment that may have water present under upset conditions?		
If site experiences sub-freezing temperatures, is there a plan for removal of ice accumulations (e.g., due to a steam leak)?		
If site experiences sub-freezing temperatures, has the lower humidity (often present with cold weather) been considered for electrostatic discharges from personnel and the hazards that may present as an ignition source?		
If site experiences heavy snow events, does the snow clearing plan consider effects of heavy snow on process equipment and/or buildings?		

If site experiences heavy snow events, does snow clearing plan consider access to or monitoring of critical process equipment/ sampling locations?		
VI. Hot Weather (Consider routine temperatures and also possible highs)		
If site experiences abnormally hot weather (for instance, 10ºC [18ºF] greater than average summer temperature), are there plans for:		
• Protecting temperature-sensitive materials?		
• Shaded or climate-controlled storage of materials normally stored outdoors?		
• Protection against thermal expansion of liquids?		
• Control of process temperatures and protection against abnormalities (e.g., exotherms, uncondensed vapors)?		
VII. Other Weather Issues		
Are there written plans for addressing weather issues, such as:		
• Drought?		
• Dust storms?		
• Forest fires?		
• Fog?		
• Frost-impact on foundations?		
• Hail		
• Sandstorms?		
VIII. Human-induced Events		
Are there written plans for addressing situations, such as:		
• Airplane, helicopter, drone, or balloon accidents?		
• Marine accidents involving ships, boats, barges, docks, flotsam, etc.?		
• Railroad accidents?		
• Car and truck accidents?		
• Accidents involving inter-company or common-carrier pipelines crossing the facility boundary?		
• Accidents involving cranes or lifting equipment?		
• Forklift accidents?		
• Major construction accidents?		
• Accidents involving dropped equipment/objects/ booms?		

• External fire (man-made) incidents?		
• Flooding (internal) incidents (e.g., failure of water storage tank, blockage of storm water drainage)?		
IX. Utility Failure (Is there a written plan to handle utility losses?)		
Air (service, plant, instrument, breathing)?		
Cooling systems (cooling water, chilled water, air, ammonia, freon, steam, nitrogen)?		
Chemical injections for antifouling or water treatment?		
Electrical power?		
Emission control?		
Flares?		
Fuels (fuel gas, natural gas, propane, bunker oil, heating oil, gasoline, kerosene, diesel)?		
Fume/dust/vapor collection?		
Heating medium (steam, electricity, hot oil, molten salt)?		
Heat tracing (steam, electric)?		
Impairment of fire water systems during maintenance activities (e.g., forbid hot work when fire water system is down for maintenance)?		
Incineration?		
Inert gas (e.g., nitrogen, argon, carbon dioxide)?		
Oil (flush, lube, seal, mineral, heat transfer)?		
Pressure/vacuum system (relief collection headers, venting, dump, quench)?		
Purge gas?		
Sewer (chemical, sanitary, storm)?		
Steam (high, medium, low pressure)?		
Ventilation/exhaust?		
Water systems, other than cooling (factory, city, well, chilled, zeolite, demineralized, cation, once-through, fire)?		
X-rays and other nuclear sources/shielding?		

INDEX